BIOFUELS, BIOENERGY AND FOOD SECURITY

Public Food Policy and Global Development Series

Series Editor: Suresh Chandra Babu

Providing expert insights from around the world into the key questions for effective policy development.
For a full list of volumes in the series visit www.elsevier.com

Public Policy and Global Development
for Food Security Series

BIOFUELS, BIOENERGY AND FOOD SECURITY

Technology, Institutions and Policies

Edited by

DEEPAYAN DEBNATH
SURESH CHANDRA BABU

ACADEMIC PRESS
An imprint of Elsevier

ELSEVIER

Academic Press is an imprint of Elsevier
125 London Wall, London EC2Y 5AS, United Kingdom
525 B Street, Suite 1650, San Diego, CA 92101, United States
50 Hampshire Street, 5th Floor, Cambridge, MA 02139, United States
The Boulevard, Langford Lane, Kidlington, Oxford OX5 1GB, United Kingdom

Notices

Knowledge and best practice in this field are constantly changing. As new research and experience broaden our
understanding, changes in research methods, professional practices, or medical treatment may become necessary.

Practitioners and researchers must always rely on their own experience and knowledge in evaluating and using
any information, methods, compounds, or experiments described herein. In using such information or methods
they should be mindful of their own safety and the safety of others, including parties for whom they have a
professional responsibility.

To the fullest extent of the law, neither the Publisher nor the authors, contributors, or editors, assume any liability
for any injury and/or damage to persons or property as a matter of products liability, negligence or otherwise, or
from any use or operation of any methods, products, instructions, or ideas contained in the material herein.

Library of Congress Cataloging-in-Publication Data
A catalog record for this book is available from the Library of Congress

British Library Cataloguing-in-Publication Data
A catalogue record for this book is available from the British Library

ISBN 978-0-12-803954-0

For information on all Academic Press publications
visit our website at https://www.elsevier.com/books-and-journals

Publisher: Charlotte Cockle
Acquisition Editor: Nancy Maragioglio
Editorial Project Manager: Devlin Person
Production Project Manager: Prem Kumar Kaliamoorthi
Cover Designer: Mark Rogers

Typeset by SPi Global, India

Dedication

Dedication of Deepayan Debnath

Dedicated to my parents, for their lifetime support, faith, and love
Nityananda and Saraswati Debnath

Dedication of Suresh Chandra Babu

Dedicated to my parents Narayanasamy Bhakthavatsalam (late)
and Govindasamy Suseela

Contents

Contributors

Nasser Ayoub
Helwan University, Cairo, Egypt

Suresh Chandra Babu
International Food Policy Research Institute, Washington, DC, United States

Miguel Carriquiry
Department of Economics, Institute of Economics, University of the Republic, Montevideo, Uruguay

Christine Costello
University of Missouri, Columbia, MO, United States

Deepayan Debnath
Food and Agricultural Policy Research Institute, University of Missouri, Columbia, MO, United States

Jerome Dumortier
School of Public and Environmental Affairs, IUPUI, Indianapolis, IN, United States

Amani Elobeid
Department of Economics and Center for Agricultural and Rural Development, Iowa State University, Ames, IA, United States

Parijat Ghosh
Food and Agricultural Policy Research Institute, University of Missouri, Columbia, MO, United States

Céline Giner
Trade and Agriculture Directorate, Organisation for Economic Co-operation and Development, Paris, France

Leila Harfuch
Agroicone, Sao Paulo, Brazil

Mike Helmar
University of Nevada-Reno, Reno, NV, United States

Marcelo M.R. Moreira
Agroicone, Sao Paulo, Brazil

Claire Palandri
School of International and Public Affairs, Columbia University, New York, NY, United States

Wyatt Thompson
Food and Agricultural Policy Research Institute, University of Missouri, Columbia, MO, United States

Simla Tokgoz
Markets, Trade and Institutions Division, International Food Policy Research Institute, Washington, DC, United States

Patrick Westhoff
Food and Agricultural Policy Research Institute, University of Missouri, Columbia, MO, United States

Jarrett Whistance
Food and Agricultural Policy Research Institute, University of Missouri, Columbia, MO, United States

Cicero Zanetti de Lima
Purdue University, West Lafayette, Indiana, USA

Preface and acknowledgments

Policy makers in both developed and developing countries need evidence to guide their choices between bioenergy and food security goals, using limited land and water resources, and in the long-run would like to transit their economies toward a bioeconomic system. The purpose of writing this book is to highlight the role of bioenergy in the context of greenhouse gas emissions mitigation in the transportation sector and its implication on food security. The first section includes an introductory chapter that describes the economic study of bioenergy systems as they relate to achieving food security and sustainable development. This chapter introduces the issues, constraints, and challenges facing the development of the bioeconomic systems that encompasses the use of natural resources to achieve sustainable development through policy institutional and technological innovations. This chapter also raises several conflicting tradeoffs between the bioenergy and food security policies that countries face in achieving sustainable development. Chapter 2 reviews the history of biofuel markets and policies and the factors that promoted the expansion of the sector. This lays the foundation for analyzing the current profile of the biofuel sector and the uncertainty surrounding its further development in the medium term. Indeed, its evolution will largely be determined by whether the possible switch in dominant conversion technologies and feedstocks effectively takes place, and the role governments might play in the energy and ecological transition.

Technological options to convert biomass to biofuel and the interactions between energy, agricultural sectors are discussed in Section II. Chapter 3 discusses alternative conversion processes from a technological perspective and identifies the opportunities for interventions to make the food system more efficient and sustainable. This discussion is concentrated on the first-generation biofuels, which is by far the most commonly used biofuels. The conversion rate from biomass to biofuel and biomass yield in terms of land are summarized. The chapter concludes with a discussion of the advantages and disadvantages of producing food crop–based biofuels. In Chapter 4, first, the interaction between the biofuels, agricultural, and energy markets is outlined; second, a new methodology is discussed in terms of linking these markets, and finally, an example of partial equilibrium model development is introduced to describe these linkages. The chapter shows the steps to quantify the impact of alternative energy price to the global agricultural commodity prices.

Section III starts with the chapter on the food-fuel-fiber debate, which summarizes the repercussions of the expansion of the first-generation biofuels, that utilizes a significant amount of agricultural feedstocks on the upstream and downstream crop value chains, livestock and dairy value chains, fiber value chains, and agricultural input value chains, affecting all consumers and producers. This chapter points out that the ensuing food-fuel-fiber debate has called into question the viability of using agricultural feedstocks for transportation sector in the long run. Cultivation of perennial cellulosic energy crop species on marginal lands, particularly in riparian areas, holds the potential to provide feedstocks for bioenergy without competing for land used for food production, while also providing a variety of ecosystem services, such as habitat for wildlife, bank stabilization, and nutrient, pesticide, and sediment attenuation that could improve water quality. These benefits are discussed in Chapter 6. Challenges and tradeoffs associated with the cultivation of these plant species on marginal lands with the target of using them as feedstocks for energy conversion are further discussed in this chapter. Chapter 7 provides a general overview of the implications of the expansion of biofuels and the related feedstock needs as sources of both direct and indirect land use changes. This chapter offers an illustration from Brazil, namely, agricultural production trends and observed land use change based on land intensification (double cropping, increased yields) and extensification (land expansion) in response to price signals. Chapter 8 evaluates emission factors for the main categories of biofuels based on a literature review and then assesses the actual emissions savings achieved by adopting the partial equilibrium Aglink-Cosimo model.

Section IV includes three chapters, which discuss alternative biofuel policy options. Chapter 9 investigates two scenarios related to biodiesel policy shifts using partial equilibrium model: (i) an increase in the US mandate for biomass-based biodiesel use; and (ii) a decrease in the allowable use of biodiesel from food-based feedstocks in the EU. The effects appear moderate, particularly for soybean and rapeseed oil prices, compared to the baseline. The effects depend, to a large extent, on model assumptions and the complex interaction of global policies and markets. The results show that these biofuel policies could have important food security implications. Chapter 10 discusses two relevant US renewable fuel policies: the Renewable Fuel Standard (RFS) at the federal level and the Low Carbon Fuel Standard (LCFS) in California. This chapter analyzes hypothetical policy regimes using a set of partial equilibrium models covering both agricultural and biofuel markets and finds that the policies are mutually reinforcing, in that the implementation of one of these policies makes meeting the other policy easier. Chapter 11 using global agricultural outlook model and geographic information system data quantify the potential increase in global commodity yields in the biofuels era.

Institutional challenges, including climate change mitigation, sustainability, and food security faced by the biofuels sectors, are discussed in Section V. Chapter 12 focuses on the definition of sustainability and discusses its importance in the context of biofuels. It concludes with the analysis of the concerns of the biofuels in the broader aspects of food security. Chapter 13 discusses the importance of supply chain management in the perspective of cellulosic biofuels. Four different optimization models are outlined in this chapter, which can assist potential biorefineries to determine optimal parcel of land and location for growing energy crops. Chapter 14, by explicitly considering the interconnection between agricultural and biofuel markets, explores the medium-term ability of the agricultural sector to supply the amount of bioenergy deemed necessary to meet greenhouse gas emissions targets in the transport sector.

We would like to thank Patrick Westhoff, Director of the Food and Agricultural Policy Research Institute, University of Missouri; Jonathan Brooks, Head of Agro-Food Trade and Markets Division, Organization for Economic Co-operation and Development; Shenggen Fan, Director General, International Food Policy Research Institute; and Rajul Pandya-Lorch, Director of Communications and Public Affairs & Chief of Staff in the Director General's Office, International Food Policy Research Institute, for their encouragement and support throughout this process.

Deepayan Debnath
Suresh Chandra Babu

SECTION 1

Introduction to bioenergy

CHAPTER 1

Bioenergy economy, food security, and development

Suresh Chandra Babu*, Deepayan Debnath[†]
*International Food Policy Research Institute, Washington, DC, United States
[†]Food and Agricultural Policy Research Institute, University of Missouri, Columbia, MO, United States

Contents

1. Introduction

In meeting their food, energy, and fuel needs, the economies of the world are slowly gearing up to move their policies and strategies toward a transformation from a food economy to a bioeconomy (EBCD, 2017; IEA, 2016; Von Braun, 2008; IEA, 2007). In this process of moving from a food economy to a bioeconomy, the immediate tradeoff between natural resource use, food security, renewable energy, waste management, and other biobased production and resource management goals becomes a key policy concern because these goals tend to compete for the same resources, such as land, labor, capital, water, and other investments (Arndt et al., 2012; IEA, 2015). For example, there has been a call from global policymakers that the development of bioenergy policies needs to be consistent with a country's overall energy security, food security, and climate change mitigation and adaptation strategies (HLPE, 2013).

Any transformation toward a bioeconomic system requires an appropriate harmonized set of policies and strategies, institutional architecture, and human capacities at the national level to address these tradeoffs (Hartley et al., 2018). Thus in terms of a national development strategy, for example, it is important to align the national bioenergy strategies to the strategies for food security, hunger reduction, poverty reduction, economic development, and conservation (Von Braun, 2008). Furthermore, sectoral policies focusing on energy, agricultural, forest management, natural resources, industry and technology, rural development, and other social sectors need to be considered when drafting bioenergy strategies (Debnath and Babu, 2018; Debnath et al., 2018). Yet, much of the emerging and developing economies are ill prepared to face these challenges due to lack of research-based evidence, institutional capacity, and human capacity to develop and implement such system-wide policies to address the growing need for interventions toward a bioeconomic transformation of the economies (Cherubini, 2010; Paloviita et al., 2017).

In addition, in this larger transformation process, the role of the bioenergy sector and biofuels in particular needs to be clearly identified (Hartley et al., 2018; Iqbal et al., 2016). All countries of the world face a variety of environmental, social, and economic challenges in the upcoming decades. The world's population is projected to increase to 8.5 billion by 2030, 9.7 billion by 2050, and exceed 11 billion in 2100 (United Nations, 2017). This would be a 28% increase in population since 2005. This increase in world population would further stress the current global food systems. A larger and more affluent population will increase the demand for essential natural resources used in the production of food, animal feed, fiber for clothing and housing materials, clean water, and energy (Dobermann and Nelson, 2013). Bioenergy development as part of and in the context of bioeconomic transformation may offer a partial solution. To address the increased demand for bioenergy, the International Energy Agency calls for an investment of about US$25–60 billion per year up to the year 2030, and about US$200 billion per year between 2050 and 2060 (IEA, 2016). Yet, it is not clear how the developing countries and the emerging economies will be able to effectively engage in the policy development process, institutional regulatory architecture, and human capacity needs to mobilize these resources and invest them to meet the growing demand for biobased energy products (Von Braun, 2008). Policymakers need to address these challenges and the tradeoffs in transforming their food economy into a bioeconomy with particular reference to bioenergy.

In the next section, we begin with a conceptual framework that helps us to understand the factors that could contribute to the process of transforming food systems into bioeconomic systems. We follow this conceptual framework in Section 3 to ask what role the bioenergy economy has to play in the development strategies and in particular how bioenergy strategies enhance the bioeconomic transformation of the economies. In Section 4 we address the tradeoff between the bioenergy and food security policies. The last section concludes.

2. A conceptual framework to identify the factors affecting the transformation of food systems into bioeconomic systems

In the last 10–15 years, particularly beginning with the 2007–08 food, fuel, and financial crises (commonly referred to as the "food crisis") the world economies faced, there has been a growing interest in understanding the implications of moving toward increased reliance on bioenergy and biofuels to reduce the pressure on increasing demand from fossil fuels. Policies and strategies that countries followed after the food crisis raised several policy and operational dilemmas. Researchers continue to understand the tradeoffs in moving toward a bioenergy strategy to manage their fuel and energy needs (Hartley et al., 2018). However, such strategic approaches have to be studied in the context of the larger transformation of the food systems that seems to be happening: a transformation of food systems from food-based economies to bioeconomies. In this section, we develop a conceptual framework for understanding the factors that speed up this process of transformation from a food economic system to a bioeconomic system.

It is well accepted that many of the current ecosystems in the world are already overexploited and unsustainable. Emerging factors and global trends such as globalization and climate change could exacerbate environmental problems by adversely affecting agricultural productivity and increasing stress on natural resource use (Baul et al., 2017; Jambor and Babu, 2016). In addition, to meet the growing food and fuel needs of the current and future population, alternatives must be explored so that countries can follow a sustainable growth path. Transforming the current food system into a bioeconomy may be a potential solution to this growing problem in the long run (European Commission, 2005, 2018).

To transform the current food systems into bioeconomic systems, we must first understand various elements of a bioeconomy. According to an early definition, the bioeconomy is the sustainable, eco-efficient transformation of renewable biological resources into food, energy, and other industrial products and uses (EBCD, 2017). Later, the bioeconomy concept was taken up by the Innovation Union for the development of the economic growth model of the European Union, recognizing that the bioeconomy represents only one but a very important sector (EBCD, 2017). A bioeconomy includes a number of characteristics and can be defined as economic growth driven by the development of renewable biological resources and biotechnologies to provide sustainable products, employment, and income (HLPE, 2013).

A bioeconomy has the potential to increase the environmentally sustainable supply of food, feed, and fiber production, improve water quality, provide renewable energy, improve the health of animals and people, and help maintain biodiversity by detecting invasive species (Oliver et al., 2014). Yet this alternative approach to growth and sustainability remains underexplored. To transform a food economy into a bioeconomy we would need at the minimum technological innovation, appropriate markets and institutions, and innovative policy processes in place. Fig. 1.1 illustrates a conceptual framework that shows how a food system can be transformed into a bioeconomic system. Based on the conceptual framework presented we provide potential steps that can be taken to carry out food system transformation, from food economy to bioeconomy.

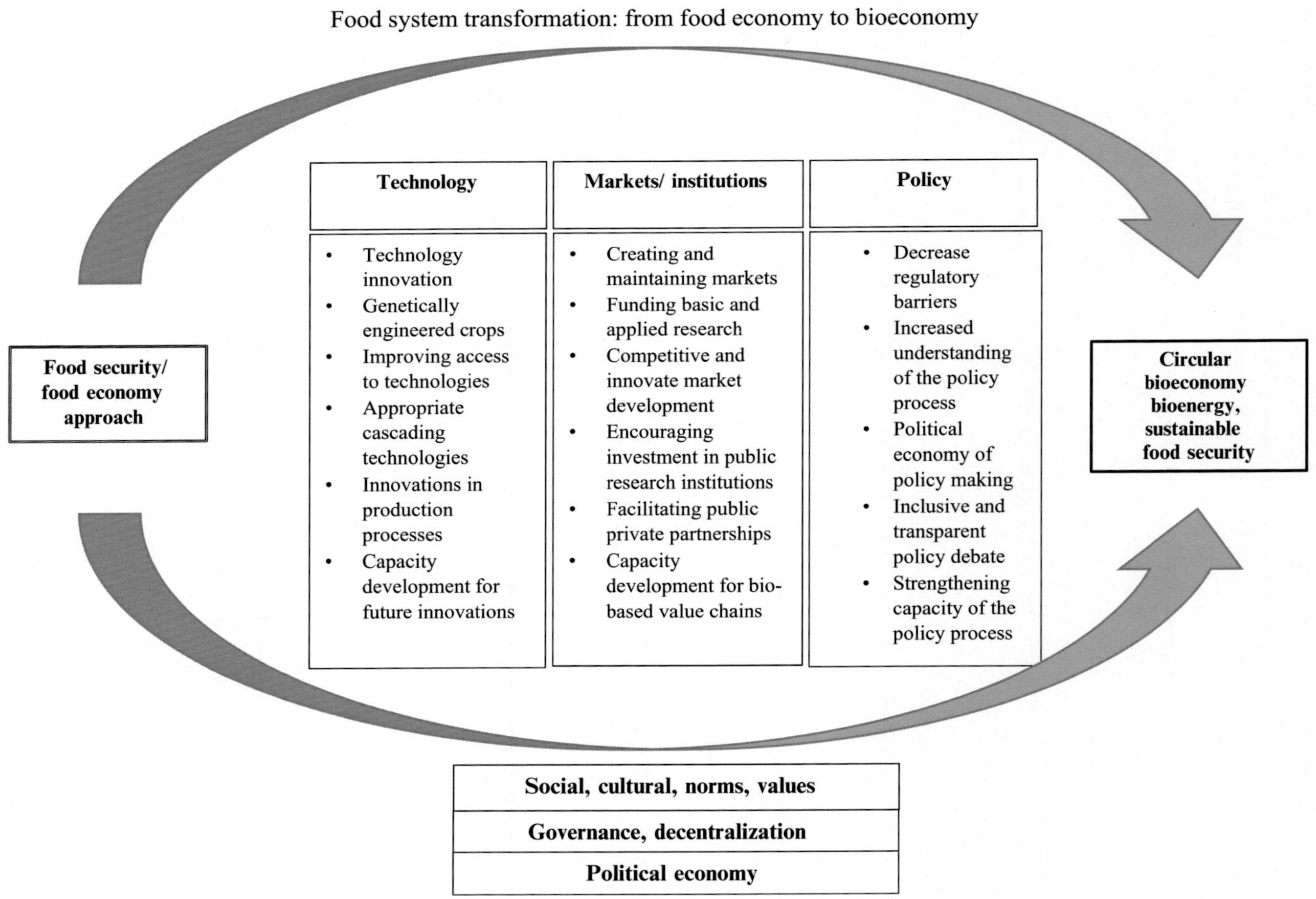

Food system transformation: from food economy to bioeconomy

Fig. 1.1 From food economy to bioeconomy.

Drivers of development of the bioeconomic system include, among other things, the increased demand for sustainable renewal of biological resources, the need to improve the management of renewable resources, and the need to respond to global challenges such as energy and food security in the face of increasing constraints on agricultural use of water, productive land use, and carbon emissions from agriculture (Priefer et al., 2017). In addition, the spread of a bioeconomic system will also depend on rapid uptake of biotechnologies in agricultural and bioenergy production, and other opportunities to decouple the growth of agriculture from environmental degradation through sustainable production methods using biotechnology (Lobell et al., 2009). The role of biotechnological advances in developing new food sources that have better access to nutrients without unduly exploiting natural resources such as land and water has been explored for some time (Babu and Rajasekaran, 1991).

As of now, many countries have published individual strategies and policies supporting biotechnology and biobased products and industries leading to bioeconomic systems. Countries are also developing strategies under one conceptual umbrella of bioeconomy instead of segregating them into individual sectors. This shift toward an advanced bioeconomy will have its spillover effects on both society in general and the environment and natural resource sustainability in particular (Dobermann and Nelson, 2013).

To ensure that the bioeconomic system reaches its full potential, we will need to identify and prepare a range of possible features to prevent locking in inferior technological solutions. To achieve this, approaches such as creating and maintaining markets for environmentally sustainable products, funding basic and applied research, and investing in multipurpose infrastructure and education will need to be combined with shorter-term policies over the next 5 years to establish a foundation for future applications. To apply this successfully in the agricultural sector, the aim is to increase the application of biotechnology to improve plant and animal varieties, improve access to technologies for producers and wholesale sellers, and expand the number of industry-driven research institutes that focus on creating such technology and systems to transform the current global food policy system. This set of issues is highlighted in the bullet points under "Technology" in Fig. 1.1.

Due to the high cost of research, regulatory barriers, and market concentration, new entrants can be prevented from hindering innovation, especially in small markets. We would need to identify potential factors that can prevent the development of a competitive and innovate market for specific biotech applications. Evaluating possible policy actions that could free up the current markets and increase access to knowledge, including encouraging public research institutions to adopt intellectual property guidelines that support rapid innovation and collaborative mechanisms for sharing knowledge, would be helpful for transforming the current markets and institutions. For the agricultural sector, for example, the slow introduction of new products due to the current regulatory environment could slow up the transformation process. Active participation of citizens and firms is required for many policies to speed up the process of transformation to a bioeconomy (Debnath and Babu, 2018). This would create an active and sustained dialog with society and industry on the socioeconomic and ethical implications, benefits, and requirements of biotechnologies (Niinimäki and Hassi, 2011). The main challenges for this food system transformation include social and institutional factors such as public opposition to adopt biotechnology, a lack of supportive regulation, and barriers to the use of biotechnology in developing countries. This set of issues is highlighted in the bullet points under "Markets/institutions" in Fig. 1.1.

Approximately half of global production of the major food, feed, and industrial feedstock crops comes from plant varieties developed using one or more types of biotechnology (IEA, 2016). This includes genetically modified crops, intragenic modification, gene shuffling, and marker-assisted selection. Researching and developing traits to improve yields and resistance to stresses such as drought, salinity, and high temperatures has increased rapidly since the early 1990s and has shown positive results in terms of increasing agricultural productivity. The time needed for a food system to transform would be determined by the extension system and policy promotions set in place. Governments have been key on accelerating the process of agrarian change, but not particularly ready to benefit from the recent rapid growth in biofuels (Priefer et al., 2017).

A transformation toward a bioeconomy is unlikely to fulfill its potential without appropriate regional, national, and in some cases global policies to support its development and implementation (Oliver et al., 2014). To set up the appropriate policies, governments will need to address misconceptions around biotechnology and describe the different alternatives for managing sustainability. The agricultural sector provides a diverse range of policy challenges that include the need to simplify regulation, encourage the use of biotechnology to improve nutritional content, ensure unhindered trade in agricultural commodities, and manage a decline in the economic viability of some sectors when faced with competition from more efficient producers. Developing effective policies that support a bioeconomy and creating environmentally sustainable alternatives are major steps for moving toward a sustained bioeconomy. A selected set of policy issues are highlighted in the bullet points under "Policy" in Fig. 1.1.

A bioeconomy has the potential to meet the needs of the growing population, decrease stress on current ecosystems from climate change, and improve and manage agricultural production substantially to 2030 (Priefer et al., 2017). The expected increase in the cost of fossil fuels over time due to declining supply of low-cost sources of petroleum, increase in demand for energy, and restrictions on the production of greenhouse gases (GHGs) could create a growing market for biomass, including nonfood crops such as grasses and trees, as a feedstock for biofuels, chemicals, and plastics (Gurria et al., 2017). Other potential markets include use of plants to produce valuable chemicals such as biopharmaceuticals and the production of nutraceuticals from plant and animal sources. These applications can be implemented depending on the investment in agricultural technology (Grassini et al., 2013).

The purpose of agriculture is more than just growing food crops. It is also growing healthy and nutritious food to create a well-nourished population (FAO, 2015). Since the agricultural sector faces interconnected challenges to ensure food security and at a time of increasing pressures from population growth, changing consumption patterns and dietary preferences, and postharvest losses, transforming the current food economy is a crucial step (Foley et al., 2011). Currently, there are opportunities and demands for several sectors of a bioeconomy such as use of biomass to provide additional renewable energy for heat, power, fuel, pharmaceuticals, and green chemicals. To meet these demands, the adoption of genetically engineered crops to increase agricultural production could be a viable solution (FAO, 2015; Mueller et al., 2012).

In summary, the transformation of food systems to a bioeconomy would be highly beneficial for current and future populations. However, the tradeoffs and the systems approach to the circular bioeconomy should be addressed in the context of its individual subsectors such as bioenergy, waste management, and biobased product development. In doing so, policy, market, institutional, and technology challenges of developing the bioenergy economy in the context of economic growth and development and the related food security and sustainability challenges should be fully studied.

3. Bioenergy strategies to move toward a bioeconomic system

Given the foregoing conceptual framework, one of the key questions to address relates to what types of bioenergy strategies are needed for a speedy transformation of food systems into bioeconomic systems. To begin with, the bioenergy strategies could be a unique entry point to the larger transformation process. The objectives that drive bioenergy strategies can be divided into two broad categories: (1) national objectives and (2) international drivers. National bioenergy program objectives, for example, can be multidirectional and depend much on the specific country's national development objectives. These are a few examples:

- Improving the income and food security of small-scale farmers through growing bioenergy crops
- Increasing farm-based, agricultural, and related jobs and the resulting rural economic development
- Increasing opportunities to improve agricultural competitiveness through introducing new bioenergy crops
- Increasing rural nonfarm activities through small- and medium-scale processing units that offer nonfarm employment and produce bioenergy and other biobased products
- Improving energy services access mainly in rural areas

- Reducing the dependency on oil imports from foreign countries to meet domestic fuel demand, which would increase national energy security
- Mitigating climate change by reducing GHG emissions by replacing fossil fuels with alternative renewable biofuels
- In land abundant countries (especially Latin American countries) improving trade balances through either exporting biofuels or agricultural feedstock required to produce it.

International drivers are basically the market demand that is driven by the biofuel policies implemented in other countries, for example, the US biofuel mandate resulted in the import of sugarcane-based ethanol from Brazil, which is counted toward the domestic advanced biofuel mandate. The biofuel policies in importing countries with respect to climate mitigation strategies and diversification in energy supply have huge consequences to exporting countries that produce biofuel and/or feedstock to supply to importing countries. Therefore the bioenergy or feedstock producers in those exporting countries have limited to no control on other countries' biofuels demands, which make them vulnerable to international bioenergy and agricultural commodities price volatility. It is the responsibility of national and local authorities in exporting countries to consider whether or not bioenergy-related investments are consistent with their country's national objectives, and assure that it does not create dependency on farmers and rural economies of the importing countries. Adequate buffering strategies are required to protect agricultural industries from international drivers in the biofuels investment process. These issues have been studied in the context of several developing countries (Arndt et al., 2010, 2012; Hartley et al., 2016; Samboko et al., 2017; Schuenemann et al., 2017; Stone et al., 2015; Thurlow et al., 2016). These studies offer several insights into policy debates on bioenergy development.

Before developing bioenergy-related strategies, set priorities, and establishing guidelines for investments in these sectors, policymakers need to consider the linkages between the various sectors and end-users. Often, the traditional agricultural and bioenergy sectors receive less attention, while the transportation and heat and power sectors receive greater attention while drafting policies (Antikainen et al., 2017). However, for bioenergy strategies to be successful, it is important to support overall economic development, food security, and poverty reduction goals as well. This requires greater emphasis on agriculture but not neglecting energy services in the rural household and small commercial sectors. Therefore efficient biofuels strategies should solve both these goals (Giljum et al., 2016).

Bioeconomic systems are increasingly linked to the circular economy in the context of how to effectively use and reuse biowastes. The bioeconomy agenda that revolves around food systems, renewable energy, waste recycling, and the production of chemicals and bioplastics could be seen as a circular bioeconomy (Paloviita et al., 2017). The circular bioeconomy also focuses on the bio value chains, biorefining, and cascading use of biomass to achieve sustainability. Researchers are concerned with the economic, sustainability, and environmental aspects of the circular bioeconomy, with particular attention given to innovations and research needs and the implications of the circular bioeconomy for social and economic transformation of societies (European Commission, 2018).

Innovation to improve productivity and to address the environmental concerns and potential use of resources are key issues to be addressed at the global level as food systems move toward circular bioeconomic systems (McMichael et al., 2007). In this context, several developing countries are trying to identify the tradeoffs and possible synergies through integrating the bioeconomy with other economic sectors. Recent developments in the United States and European Union call for a systems approach to the design and implementation of business models and policy interventions. They also call for close investigation of bioenergy policies and their impact on the agricultural, forestry, and fisheries systems on soil, water, landscape, biodiversity, and air quality in the farming communities. Increasing demand for food, fuel, and fiber will also have to be addressed through supply changes, and changes in consumer preferences, land-use patterns, and final availability of the biomass for the production of bioenergy and biobased products will have to be taken into account (Essel et al., 2014).

In this context, not only policy and institutional coordination is key for the development of the circular bioeconomy, but also the tradeoffs in the system need close attention. Investments in adequate infrastructure for the development of new products are needed. Close collaboration among various players in the bioenergy value chains is required by using a systems perspective to design policy and program interventions in the context of country-specific approaches to designing biobased interventions. Placing technology and innovation in the wider system of production, marketing, and consumption is key for successful development of the circular bioeconomy (Dammer et al., 2016).

3.1 Bioenergy in the bioeconomic system

Traditionally, the term "bioeconomy" is largely associated with the production of food for human consumption and feed for animals. Yet, other products within the bioeconomy include paper, pulp, and textiles as well as forest products, predominantly wood, which is used in construction. Local energy needs are also satisfied through the burning of wood. Within this complex global bioeconomic system, in comparison to any other renewable energy, bioenergy is much more associated with the whole system of land use, agricultural, and forestry production because it competes with other agricultural and livestock activities for land. There is now greater recognition of the potential for an expanded bioeconomy. With recent efforts to reduce GHG emissions through increasing bioenergy capacity to minimize dependence on fossil fuels and many other finite resources, demand is also increasing for the following products: recycled biobased products, added-value products sustainably produced from biomass feedstocks such as specialty chemicals derived from cellulose or lignin, wood-based building materials, and textiles. Modern and efficient energy production technologies that uses biomass are also gaining popularity.

Either new or existing bioeconomic products can provide energy and carbon savings compared to fossil-intensive products. For example, the use of wood as a building material, which can reduce or eliminate the use of steel and concrete in construction, will result in carbon sequestering for many years. Even though the actual energy and carbon benefits are debatable there are many factors, including assumptions about lifetimes and eventual disposal methods, that drive those results. In general, the use of bioeconomic materials such as wood is considered to be highly carbon efficient because it replaces materials that are products of carbon-intensive processes (Oliver et al., 2014; Kuittinen et al., 2013).

However, on the other side, the growth of the bioeconomy might lead to increased competition between the competing use of feedstocks: food, feed, materials, chemicals, and energy (Von Braun, 2008). In practice, such competition is limited because the value of bioenergy products used in energy sectors is much lower than those used for food, chemicals, or materials. In most cases, the use of biomass feedstock for energy is a small fraction of the overall energy use mainly coming from nonrenewable sources and it complements fossil resources. On the contrary, if policies and regulations are introduced to abandon fossil resources within a shorter period of time, then it could result in unacceptable socioeconomic impacts. Therefore to transit from fossil-based resources to a bioeconomy, additional policy measures might be needed. One such policy example is the introduction of carbon pricing, which applies to all affected sectors based on the magnitude of the carbon emitted or sequestered (Paloviita et al., 2017).

Effective use of bioenergy can further improve the economic and carbon benefits derived from the use of primary bioproducts. According to the theory of economies of scale, the use of bioproducts can bring down costs of new technologies, which helps to maintain existing industries' margins and strengthen the overall financial viability of new projects. A few examples are: (1) use of sawmill residues and coproducts as fuel for heating or electricity generation within the existing paper mill industry; (2) digestion of waste water and organic effluents in agro-industrial processes; and (3) integrated production of chemical products and biofuels in biorefineries. Conversely, subsidies for bioenergy production alone, without fully valuing the carbon sequestered and other benefits that can be associated with the production of those biofuels, could result in market distortions and in some cases lead to increased GHG emissions—an unintended outcome of such policies (Mueller et al., 2012).

4. Food security, food economy, and bioeconomic systems

One of the key drivers determining long-term economic viability of biofuels is competition with food. This is because biofuel production (through the use of biomass) may compete with the production of food for the same resources, notably land, labor, and water. Food security is a key developmental goal and the potential conflict with energy security can play out at many levels, including national and even regional levels. Achieving food security for all entails ensuring that all people in a country context have physical and economic access to adequate food for an active and healthy life. This food must be nutritious and safe to consume, and should also meet their dietary and cultural tastes and preferences.

Attaining food security also involves meeting its four dimensions such as availability, accessibility, utilization, and stability (FAO, 2003). In the context of biofuel development through food system transformation, the fundamental question is: To what extent could food security goals be impeded by the development of bioenergy through biofuels? Large-scale biofuel development will depend on how countries can achieve the overall balance between size of population, projected growth, availability of land (or its scarcity), and suitability for food crops versus energy crops. Other contributing factors in the development of bioenergy sources include prospects for increased productivity and the implications for land availability to meet multiple demands, as well as the relative profitability of feedstock for biofuels versus alternative uses of land, water, and labor for food, feed, or other industrial uses (FAO, 2015; Ray et al., 2013).

Incentives for feedstocks for bioenergy versus food or other crop uses will boil down to which end-product offers greater economic value added and raises the incomes of farmers, who can then afford greater access to food and nutrition (Mueller et al., 2012). When feedstocks are used for food, the availability of food will be constrained by the biofuel supply as long as they compete for the same resources (land, fertilizers, water, etc.). The impact can be more or less direct depending on the type of feedstock and where it is cultivated. There could also be indirect effects on the cropping patterns and production system of a particular crop such as in the case of the use of US-produced maize for bioethanol. Here, effects on food security might be channeled indirectly via world agricultural prices of grain and other food products whose supply and demand balances were affected by rising use of US maize in bioethanol production and use (Rosegrant et al., 2008; Schuenemann et al., 2017).

In many parts of the world, however, food access is a more critical problem than availability. Increased use of bioenergy tends to push up food prices, especially if food or feed crops are used for energy. The impact of biofuel-induced price increases will not be the same for consumers and producers. Moreover, these biofuel-induced price effects are stronger in developing countries because expenditures for food are proportionally much higher in many developing countries, and also because a large part of the population is involved in agricultural activities. Whether the net impact will be positive or negative will depend on the country, the region, and ultimately the household and individual position (Heinonen et al., 2017).

The price of food depends on the degree of processing. In developed countries, crop price has very little impact on the end food price. In developing economies, because food is less processed, higher crop prices play a greater role in setting final food prices. A higher percentage of gross domestic product is used for food, so it has a greater impact on living standards (Gerbens-Leenes et al., 2010; Ray et al., 2012).

The biofuels industry could create and improve existing market mechanisms (e.g., physical infrastructure and agronomic capability), which could lead to more efficient agricultural production. Brazil, for example, has achieved significant improvements in sugar production and ethanol processing. The by-products of biofuel production can be useful sources of food and energy. As in the United States, grain ethanol production continues to expand, and the production of dried distillers grains with solubles is expected to grow and may depress prices in the feed market. While in Brazil, with the increase in the use of sugar juice in ethanol production, bagasse (a fibrous residue by-product that remains after sugarcane stalks are crushed) is gaining popularity in terms of supplying energy (Macedo et al., 2008; Baker and Zahniser, 2006; Heinonen et al., 2017).

Another potential problem for food security is the role that biofuels might play in destabilizing food supply, especially if agricultural prices become more linked to energy prices and hence rise and fall quickly. Certainly, the tradeoffs would have to be weighed against the positive benefits (energy security, climate change, coproducts, wealth generation) and inflationary problems (Stone et al., 2015).

Overall, competition with food production is a potentially significant concern when investing in biofuels. The issue is not entirely resolved with second-generation biofuels, even if they use nonfood feedstocks because of indirect land-use changes and because of the potentially huge market demand for renewable energy in comparison to agriculture. Consequently, policies that introduce sustainability criteria and standards, if properly implemented, could contribute to mitigating this potential fuel versus food conflict (Khanna and Crago, 2012; Yijia et al., 2018; Hartley et al., 2018).

4.1 Bioenergy production: Advantages and disadvantage for food security

Bioenergy production and use, even in the context of the transformation toward a bioeconomy, has larger development-related consequences in the developing world. As a country moves toward a bioeconomy, there will be real tradeoffs in terms of the use of labor, capital, and natural resources in different types of production activities (Hartley et al., 2018). We highlight selected such tradeoffs next.

Strategies toward bioenergy development will always have environmental and socioeconomic consequences, which could be either positive or negative. It can create both opportunities and risks for food security, poverty reduction, and achieving other sustainable development goals. Reducing risks and increasing opportunities requires implementing efficient and effective policies, regulatory environment, market arrangements, and on-farm production practices to produce safe and healthy food and nonfood agricultural products that contribute to environmental, economic, and social sustainability (Mueller et al., 2012). Policy and regulatory innovations will be required at all levels of policy and decision making.

Substantial work has been done in terms of efficient and effective practices in agriculture and forestry (Kuittinen et al., 2013). However, in the case of bioenergy production there are only a few such examples where socioeconomic parameters in terms of efficient and effective practices are achieved. The following are some of the major socioeconomic dimensions that can be impacted by bioenergy production giving specific emphasis to the goal of attaining food security:

- Land accessibility
- Employment, wages, and labor conditions
- Small farmers income generation
- National food security
- Community development
- Energy security and access to energy
- Gender equity

Land accessibility

For a nation to become food secure it is crucial that its people have access to agricultural, forestry, and pasture land because it increases their ability to use the resources and ecosystem services provided by nature (Cassman, 1999; Lobell et al., 2009). Furthermore, demand for land for the development of bioenergy might trigger land-use change due to the conversion of certain agricultural and pasture lands toward the production of bioenergy feedstocks. On the one hand, this may reduce the accessibility of local communities to natural resources and can be seen as a threat to their food security. On the other hand, bioenergy development can contribute to increasing land value due to additional demand for its products coming from the bioenergy market. This may incentivize policymakers to enforce stricter land right laws to small farmers, which can be seen as potential positive effects of bioenergy development on the security of land tenure, which can further contribute to food security and poverty reduction goals (Rathmann et al., 2010).

Employment, wages, and labor conditions

The ability to purchase food by acquiring the required financial resources is another key aspect of food security, and revenue obtained through employment and wages is one of the major ways to achieve food security. Not only employment and wages but also employment quality and labor conditions are important, because the latter helps to identify whether the workers are truly benefiting from their new employment opportunities. In general, bioenergy feedstock production accounts for a significant share of total bioenergy production costs. Because investment in bioenergy development is gaining momentum in several developing countries, there are new employment opportunities along the entire supply chains, in particular, in bioenergy feedstock production. This situation is accentuated by the low level of mechanization in feedstock production operations. Wages might be positively affected if bioenergy production results in increased labor demand. However, it is important to adopt good practices for absorbing labor in bioenergy value chains, without which bioenergy production may have a negative consequence on jobs, wages, and labor conditions.

Small farmers income generation

Smallholder farmers and small and medium enterprises are self-employed and it is an important means to generate revenue that can be used to purchase food. The development of bioenergy could create new business and income-generating opportunities for them, particularly along the entire feedstock supply chain. However, they may face tough competition from larger producers because the production of bioenergy requires significant economies of scale. Therefore vertical integration in bioenergy feedstock production is inevitable, which could result in potentially excluding smallholders from the newly developed lucrative global bioenergy markets. Safeguard measures are needed to protect smallholders engaged in feedstock production through appropriate policy and regulatory measures that can help integrate them into the bioenergy production process (Ray et al., 2013; Hartley et al., 2018).

National Food Security

Feedstock demand for bioenergy may contribute to the increase in agricultural activities, including production, thereby triggering land expansion, land displacement, and/or intensification (Ray et al., 2013). The consequence is still largely unknown because some of the country-level studies assume that land displacement does not occur (Hartley et al., 2018). It may result in an increase or decrease in the supply of staple crops for food, depending on the volume of crop-based feedstock used for bioenergy production and the extent to which those crops are displaced or diverted to bioenergy production. Bioenergy feedstock production will compete with the production of staple crops for food by demanding additional resources, including land, water, labor, and fertilizers (Neumann et al., 2010). However, on the positive side, global bioenergy market development may create a number of employment and income-generating opportunities that result in increasing accessibility to food for larger sections of the rural population who are currently food insecure.

Community development

The development of bioenergy production facilities is anticipated to be in the rural areas close to agricultural fields that may provide much-needed capital investment to those areas. This may further contribute to the socioeconomic development of those local underdeveloped rural communities. It is anticipated that the local leaders of those communities may negotiate with the bioenergy companies to invest in community development programs. However, the success of these programs will depend on how well the program is designed to serve the specific needs, capacities, and desires of the targeted communities. There is high risk that a program that is poorly and ineffectively designed could lead to negative effects on local communities in terms of community wellbeing. Policymakers and local leaders have to watch for such negative effects and put in place frameworks that will help reduce their impact on community welfare (Hartley et al., 2016, 2017).

Energy security and access to energy

Energy supply security is an important aspect of development strategies in both developing and developed countries. However, developing countries that are not energy independent are more vulnerable to the demand and supply shocks in global energy markets, which may result in macroeconomic instability such as trade imbalance and inflation with certain implications on domestic food security. The production of bioenergy from domestic feedstocks diversifies the countries' energy mix and contributes to an increase in the security of energy supply, which will further contribute to a reduction in the volatility in energy access to achieve and maintain the food security of their people over time (Debnath and Babu, 2018).

Modern bioenergy developments may increase energy accessibility, particularly in rural areas with positive effects on food security. Access to energy also contributes to food security in two ways: (1) it increases the productivity of the agricultural sector resulting in higher food production and availability, and (2) the use of energy services for cooking is important for food preparation and utilization.

Within rural communities, bioenergy might replace traditional energy sources such as charcoal, which reduce air pollution and have positive health effects as well (Mantau, 2012).

Gender equity

The consequences of commercial agriculture on gender inequity have been studied for some time. The production of bioenergy feedstocks by smallholder farmers will increase the commercialization of this sector. Like any other industry, gender-based inequalities can persist in the newly developed bioenergy sector. In most developing countries, access to and control over resources and assets (such as land, water, and other natural resources; agricultural inputs and equipment; agricultural extension services; credit, particularly formal credit schemes; and market access) are only given to male members of households. If these bioenergy feedstock production and related production units continue to exclude women from their benefits, then there is the risk that the modern bioenergy industry would fail to achieve gender equity, which might further affect the food security of those households. Understanding and learning how other commercial value chains have addressed such gender equity issues would be paramount to the development of the bioenergy industry (FAO, 2015; European Commission, 2018).

5. Concluding remarks

This chapter aimed to raise some of the pressing issues facing the interface of bioenergy economy, food security, and the border development process. The issues were discussed in the context of moving from a food system approach to development to bioenergy within the global bioeconomic system. The chapters of this book are intended to provide up-to-date knowledge of the current state of the countries that are in the process of transforming from a food system economy to a bioeconomic system. They review the current state of and the implications of the existing bioenergy and biofuel policies. In the quest to become energy independent, countries are currently searching for possible long-term strategies and are looking at opportunities for developing internal debates and dialogs in their policy systems. The chapters of this book are likely to benefit the student enrolled in bioenergy and food security-related courses, as well as prospective researchers in this emerging area in economic growth and development processes and in the structural transformation of economies.

Some of the major issues that are open to research, evidence generation, and policy dialog are as follows. Given that the emerging economies and developing countries are increasingly facing the growing need for energy to speed up their economic growth and development, new sources of energy are being continuously explored. Bioenergy development and use have a high potential to address the growing demand for energy. Along with solar energy and wind energy, bioenergy sources are supposed to help countries meet their energy demands (European Commission, 2018; IEA, 2015). Bioenergy sources are part of the low-carbon alternatives that meet global carbon mitigation commitments. They could help in decarbonizing the transportation sector, particularly in the long road, shipping, and aviation sectors. Bioenergy sources are expected to provide about 17% of total energy demand in 2060 (IEA, 2016).

At present, close to 90% of bioenergy production and use is confined to countries and regions such as China, Brazil, the European Union, and the United States. However, in the years to come it is expected that bioenergy sources will play a key role in the larger development of the bioeconomic system. Yet, there are several challenges that should be addressed to make this transformation. There is currently low investment in research and development leading to the development of appropriate technologies. The production, conversion, and utilization of bioenergy products continue to be a complex process requiring the overcoming of several policy and regulatory hurdles. Finally, the link to the broader bioeconomy has not been fully understood by policymakers at the global, regional, and national levels (Hartley et al., 2018).

In this context the chapters of this book ask what policy, institutional, and technology lessons have been learned so far from the countries that have adopted bioenergy policies and implemented them successfully. This book also explores the opportunities, challenges, and potential solutions that could be explored in the development of the bioenergy sectors and linking them to countries' larger bioeconomic systems. Developing countries need to diversify farm production, and bioenergy crops could be an effective way to reduce poverty and at the same time address local energy needs, which for the most part currently depend on fossil fuels. Farmers could earn cash by growing bioenergy crops and reduce their food insecurity levels substantially. Yet, the conflict between the land uses for various competing crops will require policy and institutional coordination at all levels (Arndt et al., 2012).

Risk-averse smallholder farmers may not move toward bioenergy crops unless they have an assured market for their production. Several supporting mechanisms should be in place to make the transition from a food system to a bioeconomic system in the context of developing countries. The development and use of technology should be toward identifying opportunities for bioenergy crops in the production of biomethane, fuel for heating systems, generation of local electricity, and effectively using biofuels in transportation systems. This will further call for identifying the facilitating markets and market access to farmers, providing low-cost financing to small and medium enterprises to set up and operate biofuel units and increase the technical and institutional capacity throughout the bioenergy value chain (Arndt et al., 2010; Babu and Hallam, 1989).

As countries explore the opportunities to transform their food system toward bioeconomic systems and address their increasing energy demands in a sustainable manner, the role of bioenergy in meeting food security and poverty reduction goals becomes paramount. In this process the possible adverse effects of growing and using biomass from land previously used for food production on food security goals needs to be monitored. The issue of environmental sustainability and the effective use of existing biofuel value chains requires policy attention in developing countries. Finally, the role of increased investment in research and innovation toward developing sustainable technologies that can reduce risks associated with bioeconomic transformation cannot be underestimated (Gonzales et al., 2013).

References

Antikainen, R., Dalhammar, C., Hildén, M., Judl, J., Jääskeläinen, T., Kautto, P., Koskela, S., Kuisma, M., Lazarevic, D., Mäenpää, I., Ovaska, J.-P., Peck, P., Rodhe, H., Temmes, A., Thidell, Å., 2017. Renewal of Forest Based Manufacturing Towards a Sustainable Circular Bioeconomy. Finnish Environment Institute, Helsinki.

Arndt, C., Benfica, R., Tarp, F., Thurlow, J., Uaiene, R., 2010. Biofuels, poverty, and growth: a computable general equilibrium analysis of Mozambique. Environ. Dev. Econ. 15 (1), 81–105.

Arndt, C., Pauw, K., Thurlow, J., 2012. Biofuels and economic development: a computable general equilibrium analysis for Tanzania. Energy Econ. 34, 1922–1930.

Babu, S.C., Hallam, J.A., 1989. Evaluating agricultural energy policies under uncertainty—the case of electricity in South India. Agric. Econ. 3, 187–198.

Babu, S.C., Rajasekaran, B., 1991. Biotechnology for rural nutrition: an economic evaluation of algal protein supplements in South India. Food Policy 16 (5), 405–414.

Baker, A., Zahniser, S., 2006. Ethanol reshapes the corn market. USDA ERS Amber Waves 4(2).

Baul, T., Alam, A., Ikonen, A., Strandman, H., Asikainen, A., Peltola, H., Kilpeläinen, A., 2017. Climate change mitigation potential in boreal forests: impacts of management, harvest intensity and use of forest biomass to substitute fossil resources. Forests 8 (12), 455. https://doi.org/10.3390/f8110455.

Cassman, K.G., 1999. Ecological intensification of cereal production systems: yield potential, soil quality, and precision agriculture. Proc. Natl. Acad. Sci. U. S. A. 96, 5952–5959.

Cherubini, F., 2010. The biorefinery concept: using biomass instead of oil for producing energy and chemicals. Energy Convers. Manag. 51 (7), 1412–1421. https://doi.org/10.1016/j.enconman.2010.01.015.

Dammer, L., Bowyer, C., Breitmayer, E., Nanni, S., Allen, 2016. The circular economy and the bioeconomy—partners in sustainability 55 B. In: Carus, M., Essel, R. (Eds.), Cascading use of wood products. World Wide Fund for Nature.

Debnath, D., Babu, S.C., Ghosh, P., Helmer, M., 2018. The impact of India's food security policy on domestic and international rice market. J. Policy Model 40 (2), 265–283. https://doi.org/10.1016/j.jpolmod.2017.08.006.

Debnath, D., Babu, S.C., 2018. What does it take to stabilize India's sugar market? Econ. Polit. Wkly 53 (36), 16–18. https://www.epw.in/journal/2018/36/commentary/what-does-it-take-stabilise-indias-sugar.html.

Dobermann, A., Nelson, R., 2013. Opportunities and Solutions for Sustainable Food Production. United Nations Sustainable Development Solutions Network, New York, NY.

EBCD, (European Bureau for Conservation and Development), 2017. The Role of Bioeconomy in the Circular Economy. European Bank for Construction and Development, Brussels, Belgium.

Essel, R., Breitmayer, E., Carus, M., Fehrenbach, H., von Geibler, J., Bienge, K., Baur, F., 2014. Discussion Paper: Defining Cascading Use of Biomass. Nova-Institut GmbH, Hürth, Germany.

European Commission, 2005. New Perspectives on the Knowledge Based Bio-Economy: A Conference Report. European Commission, Brussels, Belgium.

European Commission, 2018. Communication From the Commission to the European Parliament, the Council, the European Economic and Social Committee and the Committee of the Regions on a Monitoring Framework for the Circular Economy, COM (2018). p. 29 (final).

FAO, 2015. Food Wastage Footprint & Climate Change. Food and Agriculture Organization of the United Nations, Rome.

FAO (Food and Agricultural Organization of the United Nations), 2003. Trade Reforms and Food Security. FAO, Rome.

Foley, J.A., Ramankutty, N., Brauman, K.A., Cassidy, E.S., Gerber, J.S., Johnston, M., Mueller, N.D., O'Connell, C., Ray, D.K., West, P.C., Balzer, C., Bennett, E.M., Carpenter, S.R., Hill, J., Monfreda, C., Polasky, S., Rockström, J., Sheehan, J., Siebert, S., et al., 2011. Solutions for a cultivated planet. Nature 478 (7369), 337–342. https://doi.org/10.1038/nature10452.

Gerbens-Leenes, P.W., Nonhebel, S., Krol, M.S., 2010. Food consumption patterns and economic growth. Increasing affluence and the use of natural resources. Appetite 55, 597–608.

Giljum, S., Bruckner, M., Gözet, B., de Schutter, L., 2016. Land Under Pressure: Global Impacts of the EU Bioeconomy. Friends of the Earth Europe, Brussels.

Gonzales, C., Trigo, E., Herrera Estrella, L., Farias, A., 2013. Current Status and Future Potential of Knowledge Based Bio-Economy Related Research & Innovation in Latin America and the Caribbean and Policy Recommendations. (Bioeconomy Working Paper No. 2013-02).

Grassini, P., Eskridge, K., Cassman, K., 2013. Distinguishing between yield advances and yield plateaus in historical crop production trends. Nat. Commun. 4, 2918.

Gurria, P., Ronzon, T., Tamosiunas, S., Lopez, R., Garcia Condado, S., Guillen, J., Cazzaniga, N.E., Jonsson, R., Banja, M., Fiore, G., Camia, A., M'Barek, R., 2017. Biomass Flows in the European Union. JRC technical reports, Joint Research Centre, Ispra, Italy.

Hartley, F., van Seventer, D., Tostao, E., Arndt, C., 2016. Economic Impacts of Developing a Biofuel Industry in Mozambique. WIDER Working Paper 2016/177, United Nations University, Helsinki.

Hartley, F., van Seventer, D., Samboko, P.C., Arndt, C., 2017. Economy-Wide Implications of Biofuel Production in Zambia. WIDER Working Paper 2017/27, United Nations University, Helsinki.

Hartley, F., van Seventer, D., Sanmboko, P.C., Arndt, C., 2018. Economy-wide implications of biofuel production. Dev. South. Afr., 1–20.

Heinonen, T., Pukkala, T., Mehtätalo, L., Asikainen, A., Kangas, J., Peltola, H., 2017. Scenario analyses for the effects of harvesting intensity on development of forest resources, timber supply, carbon balance and biodiversity of Finnish forestry. Forest Policy Econ. 80, 80–98.

HLPE, 2013. Biofuels and Food Security—A Report by the References the Circular Economy and the Bioeconomy—Partners in Sustainability 57 High Level Panel of Experts on Food Security and Nutrition of the Committee on World Food Security. Food and Agriculture Organization of the United Nations, Rome.

IEA, 2007. IEA bioenergy task 42 on biorefineries: coproduction of fuels, chemicals, power and materials from biomass. In: Minutes of the Third Task Meeting, Copenhagen, Denmark, 25–26 March 2007. International Energy Agency, Paris.

IEA, 2015. Nutrient Recovery by Biogas Digestate Processing. International Energy Agency, Paris. IEA, 2016, Key world energy statistics, International Energy Agency. IEA Bioenergy, 2014a, Annual report 2014 IEA Bioenergy.

IEA, 2016. Tracking Clean Energy Progress 2016. Energy Technology Perspectives 2016 Excerpt. IEA Input to the Clean Energy Ministerial. International Energy Agency, Paris, France.

Iqbal, Y., Lewandowski, I., Weinreich, A., Wippel, B., Pforte, B., Hadai, O., Tryboi, O., Spöttle, M., Peters, D., 2016. Maximising the Yield of Biomass From Residues of Agricultural Crops and Biomass From Forestry. Final Report, Ecofys, Berlin.

Jambor, A., Babu, S.C., 2016. Agricultural Competitiveness and Global Food Security. Springer, New York.

Khanna, M., Crago, C.L., 2012. Measuring indirect land use change with biofuels: implications for policy. Ann. Rev. Resour. Econ. 4 (1), 161–184.

Kuittinen, M., Ludvig, A., Weiss, G., 2013. Wood in Carbon Efficient Construction: Tools, Methods and Applications. CEI-Bois.

Lobell, D.B., Cassman, K.G., Field, C., 2009. Crop yield gaps: their importance, magnitudes, and causes. Annu. Rev. Environ. Resour. 34, 179–204.

Macedo, I.C., Seabra, J.E.A., Silva, J.E.A.R., 2008. Green house gases emissions in the production and use of ethanol from sugarcane in Brazil: the 2005/2006 averages and a prediction for 2020. Biomass Bioenergy 32, 582–595.

Mantau, U., 2012. Wood flows in Europe (EU27), project report. In: Confederation of European Paper Industries and European Confederation of Woodworking Industries, Celle, Germany.

McMichael, A.J., Powles, J.W., Butler, C.D., Uauy, R., 2007. Food, livestock production, energy, the circular economy and the bioeconomy—partners in sustainability climate change, and health. Lancet 370 (9594), 1253–1263. https://doi.org/10.1016/S0140-6736(07)61256-2.

Mueller, N., Gerber, J., Johnston, M., Ray, D., Ramankutty, N., et al., 2012. Closing yield gaps: nutrient and water management to boost crop production. Nature 490, 254–257.

Neumann, K., Verburg, P., Stehfest, E., Müller, C., 2010. The yield gap of global grain production: a spatial analysis. Agric. Syst. 103, 316–326.

Niinimäki, K., Hassi, L., 2011. Emerging design strategies in sustainable production and consumption of textiles and clothing. J. Clean. Prod. 19 (16), 1876–1883.

Oliver, C.D., Nassar, N.T., Lippke, B.R., McCarter, J.B., 2014. Carbon, fossil fuel, and biodiversity mitigation with wood and forests. J. Sustain. For. 33 (3), 248–275. https://doi.org/10.1080/10549811.2013.839386.

Paloviita, A., Kortetmäki, T., Puupponen, A., Silvasti, T., 2017. Insights into food system exposure, coping capacity and adaptive capacity. Br. Food J. 119 (12), 2851–2862.

Priefer, C., Jörissen, J., Frör, O., 2017. Pathways to shape the bioeconomy. Resources 6 (1), 10.

Rathmann, R., Szklo, A., Schaeffer, R., 2010. Land use competition for production of food and liquid biofuels: an analysis of the arguments in the current debate. Renew. Energy 35 (1), 14–22.

Ray, D.K., Ramankutty, N., Mueller, N.D., West, P.C., Foley, J.A., 2012. Recent patterns of crop yield growth and stagnation. Nat. Commun. 3, 1293.

Ray, D.K., Mueller, N.D., West, P.C., Foley, J.A., 2013. Yield trends are insufficient to double global crop production by 2050. PLoS ONE 8(6).

Rosegrant, M.W., Zhu, T., Msangi, S., Sulser, T., 2008. Global scenarios for biofuels: Impacts and implications for food security. Rev. Agric. Econ. 30 (3), 495–505.

Samboko, P.C., Subakanya, M., Dlamini, C., 2017. Potential Biofuel Feedstocks and Production in Zambia. WIDER Working Paper 2017/47, United National University, Helsinki.

Schuenemann, F., Thurlow, J., Zeller, M., 2017. Leveling the field for biofuels: comparing the economic and environmental impacts of biofuel and other export crops in Malawi. Agric. Econ. 48, 301–315. https://doi.org/10.1111/agec.12335.

Stone, A., Henley, G., Maseela, T., 2015. Modelling Growth Scenarios for Biofuels in South Africa's Transport Sector. WIDER Working Paper 2015/148, United Nations University, Helsinki.

Thurlow, J., Branca, G., Felix, E., Maltsoglou, I., Rincón, L.E., 2016. Producing biofuels in low-income countries: an integrated environmental and economic assessment for Tanzania. Environ. Resour. Econ. 64 (2), 153–171.

United Nations, 2017. World Population Prospects: The 2017 Revision. 21 June 2017, United Nations, New York.

Von Braun, J., 2008. The Role of Science and Research for Development Policy and the Millennium Development Goals. Humboldt Foundation, Berlin.

Yijia, L., Miao, R., Khanna, M., 2018. Effects of ethanol plant proximity and crop prices on land-use change in the United States. Am. J. Agric. Econ., aay080. https://doi.org/10.1093/ajae/aay080.

CHAPTER 2

Technology, policy, and institutional options

Claire Palandri*, Céline Giner†, Deepayan Debnath‡
*School of International and Public Affairs, Columbia University, New York, NY, United States
†Trade and Agriculture Directorate, Organisation for Economic Co-operation and Development, Paris, France
‡Food and Agricultural Policy Research Institute, University of Missouri, Columbia, MO, United States

Contents

Acronyms

bln	Billion
BtL–FT fuels	Biomass to Liquids fuels. They are one of the three categories of synthetic diesel fuel or "synfuels," by which is meant the product made by Fischer-Tropsch (FT) synthesis from syngas. Synfuels that are produced from biomass—in comparison to coal or gas—are often referred to as BtL-FT fuels
coe	Crude oil equivalent
DME	Dimethyl ether. Is a substitute for diesel fuel
ETBE	Ethyl tertiary butyl ether. It is a biofuel derived from ethanol (47% v/v) and isobutylene (53% v/v), which can be blended with gasoline
FAEE	Fatty acid ethyl ester
FAME	Fatty acid methyl ester (the generic chemical term for biodiesel)
gal	Gallons
GHG	Greenhouse gas
HEFA/HVO	Hydro-processed esters and fatty acid, also referred to as hydrotreated vegetable oils
HPO	Hydrotreated pyrolysis oils
LNG	Liquefied natural gas
MSW	Municipal solid waste
Mt	Megatons
MTG	Methanol-to-gasoline
PPO	Pure plant oils

Biofuels, Bioenergy and Food Security
https://doi.org/10.1016/B978-0-12-803954-0.00002-4

RFS   Renewable Fuel Standard
SNG   Substitute natural gas
STG   Syngas-to-gasoline

1. Introduction

Since the 1970s, countries around the world have started to implement policies promoting renewable sources of energy as a viable alternative to the traditionally used fossil fuels in the transportation sector. They used multiple policy instruments and development programs in order to increase the production and consumption of biofuels. Biofuel markets are mostly policy driven and have emerged in different countries as a way to achieve distinct objectives. Policies implemented in developed as well as in developing countries were motivated by different considerations, leading to the multiplication of various economic instruments. Moreover, the biofuels sector is fairly complex as it constitutes a particular commodity group that lies at the intersection of the agricultural and energy domains: two sectors involving different actors, value chains, and concerned with diverse issues.

Once bioenergy policies have been successfully designed, institutions play a critical role in ensuring their implementation and enforcement. The link between economics and policies and the physics of biofuels goes both ways: the development of the biofuels industry greatly depends on policies implemented, incentives introduced, and financial support. Reciprocally, the development of new types of biofuels will impact the size of biofuels markets, and the hierarchy of producing countries. Economic viability is a permanent concern, and the coming years will likely be decisive for the direction of biofuel markets, as new technologies should become price competitive. However, the high dependency on domestic and trade policies and uncertainties concerning their evolution make it difficult to precisely project the evolution of biofuel markets.

Technology considerations are also key to understanding the current state of the biofuels industry and the evolution of consumption and production patterns. Indeed, biofuels have proved to be a very diverse market, for which production technologies are evolving rapidly. While some second-generation biofuels[1] are reaching commercialization, many still remain at the research and development stage, which increases the level of uncertainty regarding the future of the market. The evolution of newer and more cost-effective technology—and satisfying recent and increasingly ambitious sustainability objectives—will probably be decisive in the development of new types of biofuels.

For these reasons, this chapter will address both economic and technological concerns by exposing a global picture of biofuel policies and the current state of technology. Analyzing the current situation of markets and how they evolved to the way they are today is critical to understanding how governments can build on existing instruments to achieve current objectives.

---

[1] "Second-generation" biofuels generally refer to the transformation of waste, residues, or cellulosic material. They are discussed further in the following pages.

This chapter is organized as follows. The first section provides a first appreciation of the biofuels industry, introduces and defines the main categories of biofuels, and presents a snapshot of the current state of biofuel markets. The second section provides an extensive overview of the evolution of biofuel policies throughout the world by describing the major policy reforms since the beginning of the development of the industry—with a focus on recent developments. The last section focuses on technology and discusses the major pathways for the conversion of feedstocks into different categories of biofuels. All numbers presented in the chapter to describe biofuel markets are based on the 2017 OECD-FAO Agricultural Outlook database (OECD-FAO, 2017).

1.1 A brief introduction to biofuels

As their name indicates, biofuels refer to fuels produced by the transformation of biomass; they are thus "biomass based." This all-encompassing definition includes both liquid and gaseous biofuels; however, in this chapter we will focus on the first category.

Biofuels constitute a source of renewable energy that can be used as transportation fuels, either directly or by being blended in conventional fossil fuel pools. A first definition would thus be that they are substitutes for fossil fuels. In 2015, biofuels used in the transportation sector represented about 4% of global road transport fuel (IEA, 2016). In the scope of road transportation,[2] two main types of biofuels are distinguished: gasoline substitutes and diesel substitutes. Generally speaking, bioethanol—hereafter referred to as ethanol—is a substitute for gasoline and biodiesel a substitute for diesel.

Because of their molecular structure composed mostly of hydrocarbon chains, their combustion leads—similarly to fossil fuels—to the emission of sizeable volumes of carbon dioxide (CO_2).

The biofuels industry is relatively recent, with volumes produced and consumed becoming significant only in the 1990s, and remains concentrated in a small number of countries. In 2015, 12 countries shared 97% of the global amount of biodiesel consumed, and 11 countries consumed 97% of ethanol produced for fuel use, while global production volumes averaged at about 115,600 million liters of ethanol and 30,900 million liters of biodiesel. The United States and Brazil dominate the ethanol market, accounting, respectively, for 49% and 25% of global production. Biodiesel production is mostly driven by the European Union, which is responsible for 39% of global volumes.

The first biofuels developed—which still make up the vast majority of global production—were derived from agricultural crops. They are commonly referred to as "conventional" or "first-generation" biofuels and can originate from various feedstocks.

---

[2] The potential of biomass-based fuels for the transport sector is not limited to road vehicles. For advanced biofuels especially, biofuels are making their way in the aviation sector as well, synthesized in the form of drop-in fuels. However, these technologies are still at the experimental stage. For this reason, this report focuses on the use of biofuels in road vehicles.

The most common include grains (maize, wheat), sugar cane, and cassava for bioethanol production, and palm oil, soybean oil, and jatropha for biodiesel production.

An important distinction is made between these conventional biofuels and the more recent "advanced" or "second-generation" biofuels, which refer to the transformation of waste, residues, or cellulosic material. However, no singular and internationally accepted definition exists to this day. This is in part due to the diversity of feedstocks and technologies of production, which are very different among countries and make it difficult to classify biofuels. Technologies for the production of advanced biofuels are fairly recent and this sector has not reached an industrial scale. Finally, "third-generation biofuels" correspond to the production of biofuels from algae. We will not be addressing these in this chapter.

In the next section we conduct a review of biofuel policies, which enables us to understand how this newly emerged market has attained its current situation (it is also the basis on which to reflect where we go from here).

2. Overview of the evolution of biofuel policies and institutional options

2.1 1970s–2000s

Before the 2000s, the biofuels industry was present in a very limited number of countries and most of the production was located in Brazil. The utilization of ethanol as a transport fuel existed in several developed and developing countries since the 1920s; however, it remained limited and was not regulated by substantial policies. The use of sugar cane for the production of fuel ethanol in Brazil became significant in the 1970s.

Throughout that period, the development of the biofuels industry was driven mainly by international oil prices and was subjected to its variations. For as long as oil prices were relatively high, the biofuels industry developed significantly in several countries. On the contrary, periods of low international crude oil prices decreased the competitiveness of biofuels.

South Africa used bioethanol from sugar cane in petrol from the 1920s until the 1960s. Because it was neither supported by policies nor attracted sufficient investment, when prices dropped in the 1960s, ethanol was no longer competitive and its consumption dropped. The lack of support and policies led to a stagnation of the development of the industry. Although the 1973 and 1979 oil crises did not reverse the trend, the later high oil prices that occurred around 2005 and climate change considerations led to major national and international interest in investing in biofuels production again. The first effective biofuel policies in South Africa were implemented at the beginning of the 2000s.

In Brazil, the biofuels sector became significant after 1973, when the crude oil price on the international market jumped from $2.7 to $11 per barrel. In response to the oil crisis, the national "Pró-Álcool" program was launched by the Brazilian government in

1975 as a way to reduce oil imports for gasoline production by favoring domestic production of ethanol. It introduced the first national biofuel policies, with instruments and measures on both the demand and the supply sides. The government provided support through production subsidies and granted tax exemptions to ethanol fuel sales and purchases of ethanol-powered cars. It also mandated the installment of ethanol pumps and set at the forefront a mandatory blending[3] of ethanol with gasoline of 4.5%.

The second oil crisis of 1979 initiated the beginning of the second phase of the Pró-Álcool program, which increased the blending mandate to 22% and introduced a producer price fixing scheme. The government introduced subsidies to guarantee lower prices at the pump with a 65% price ratio between hydrated ethanol (E100) and gasoline prices. Until the 1990s, the sector remained heavily regulated. In 1990, production reached about 11,800 million liters at which point ethanol fuel consumption represented approximately 50% of total light vehicle fuel consumption.

In the 1990s, however, the industry faced both a drop in petroleum prices (the "petroleum countershock") and an increase in the world sugar price, leading to a shortage of ethanol production. In addition to matters of energy independence, the sector faced issues related to biofuels competition with food supplies, which became a new major concern. The 1989–90 ethanol supply crisis began the road to ethanol deregulation and the end of the Pró-Álcool program. Ethanol prices were finally liberalized by 1999 and the distribution monopoly was eliminated, following the deregulation of sugar prices. The government's direct control over the volume of sugar or ethanol that could be produced, consumed domestically, or exported was also abolished and hydrous ethanol subsidies were eliminated. Ethanol production and price-setting schemes have since been influenced solely through the ethanol-use mandate and tax incentive measures.

Before the 2000s, tax exemptions and production subsidies for biofuels were also introduced in several developed countries. The United States was one of the first OECD member countries to exempt ethanol-petrol blends from a fuel excise tax, as it implemented a total exemption from the motor fuels excise tax for ethanol in 1979. In Australia, ethanol was granted an excise tax exemption in 1980. In Canada, the ethanol portion of blended gasoline was exempted from the federal excise tax in 1992. These initial policies were replaced by more complex and precise measures at the beginning of the 2000s. In the United States, the support provided through total tax exemption engendered significant revenue losses; it was then replaced in 2004 by more specific tax reductions. Production subsidies were also implemented in the 1990s to foster the scale-up of the

---

[3] Blending or use mandates are obligations to blend certain proportions of biofuels with fossil fuels, or binding targets for shares of biofuels in energy use. The overwhelming majority of mandate levels are set on a volume basis. In the European Union, however, levels are set according to the energy content of fuels, and in Germany they are based on greenhouse gas (GHG) emissions savings. Mandates represent a significant subsidy to biofuels producers, who are guaranteed a market regardless of the price of fuel.

industry. While credits were allocated for small biofuels producers in the United States in 1990, Australia introduced bounties from 1994 to 1997, designed to help new ethanol producers achieve commercial production.

However, it was not until the beginning of the 2000s that rigorous plans for the development of the biofuels industry and resulting specific programs and measures were implemented.

2.2 2002–2010

A surge originating from various concerns

In the 2000s, the biofuels industry expanded rapidly to several developed and developing countries, driven by multiple support policies. Policies were motivated by various concerns, in part as the result of the sector's particular position at the crossroads between energy and agricultural markets. A concern for many was energy security, and the development of biofuels seemed to be a possible solution toward the reduction of energy imports. This was particularly the case of countries that rely heavily on foreign fuels, such as Japan. Developing countries such as Argentina and South Africa promoted the biofuels industry almost exclusively to enhance rural development, create jobs, and generate economic activity (Republic of South Africa, 2007). In developed countries, biofuels were seen as a way to reduce greenhouse gas (GHG) emissions, in comparison with the burning of fossil fuels. Finally, other countries implemented policies to address issues in domestic agricultural markets: China administered its first fuel ethanol policies and programs in the early 2000s in response to an abundance of grain production, and the promotion of the biofuels industry in the European Union aimed at supporting food prices in a context of low international prices.

The early phase: Pilot programs

In the early 2000s, several pilot projects were launched in Asian countries in which the industry was nascent. Most were designed to test the implementation of a blending mandate or analyze infrastructure needs, and were restricted to a few cities or provinces, with the ambition to be increased or expanded in the near future.

In 2001, China and India launched ethanol promotion programs in a selection of cities and territories. A similar project was launched in South Korea in 2002 to test the viability of a 20% biodiesel blend, followed a few years later by a study on ethanol; both focused notably on the need for infrastructure. As a result of these pilot projects, the use of biofuels was gradually introduced in larger portions of these countries and led to the creation of national biofuel strategies and wider support policies. India initiated its Ethanol Blending Program in 2003, imposing a 5% blending mandate gradually in most states.

The multiplication of support policies

The vast range of initial concerns and factors for the implementation of promotion measures led to a proliferation of support policies and associated instruments in the 2000s. The United States was the first developed country to develop specific instruments, immediately followed by a number of governments around the world. Since the early 2000s, the production and consumption of biofuels have remained mostly policy driven.

A wide array of different instruments and incentives were introduced in the first stages of the development of domestic industries. A majority of countries implemented partial or full exemptions of excise taxes. In 2004, the European Union's first policy on biofuels authorized member states to adopt a specific taxation system. Several governments also introduced blending mandates while aiming to increase them over time. Among many indirect support policies, countries also implemented preferential loans or tax incentives for "clean vehicles." In Brazil, flex-fuel cars were launched in March 2003 and were granted tax reductions the following year.

In particular, the biofuels sector was strongly regulated in the first few years in many emerging economies. A few, such as Argentina and India, created price-fixing schemes. Since the National Policy on Biofuels of 2009, Indian biofuels suppliers can only sell their supplies to oil marketing companies (OMCs) that purchase it at a minimum purchase price set by the government. In Argentina, supply prices have been fixed since 2008 for biodiesel and 2012 for ethanol. Since 2005, biodiesel production in Brazil has been supported through tax reductions, blending mandates, and a public auction system that sets and allocates the total production volume and the opening price, while the ethanol sector, which represents 89% of Brazil's biofuels industry, is supported through a single system of flexible blending mandates. Since 2003, the Brazilian government has set the minimum percentage of the ethanol blend according to the results of the sugar cane harvest and the levels of ethanol production from this feedstock, resulting in frequent blending variations. Finally, some countries such as South Africa see government intervention and regulation as a transition phase toward a decrease in incentives over time (Republic of South Africa, 2007). In China and India, however, ethanol development still remains heavily regulated.

Export opportunities for developing countries

As a result of the setting of biofuel targets in developed countries—mostly the European Union and the United States—a surge in production occurred in several developing countries, which saw in it a great export opportunity. Multinational companies became interested in large-scale production in these countries and sought to acquire large tracts of land to establish plantations. In Mozambique, declarations made during the 2004 election campaign, promoting jatropha cultivation on marginal lands to produce biofuels, led to huge investments from local and foreign private investors (notably Petrobras) up to 2008. Similarly, a flow of investors and companies entered Tanzania from 2005 onward and invested in the country's biofuels industry, interested in large-scale production of biofuels to export to the European Union. However, the acquisition of land in countries without any policy or framework in place ultimately led to a severe decline in investments and abandonment of many projects and operating plants.

The purpose of production for exportation also led developing countries to implement biofuel policies in regard to developed countries' standards. The South African Biofuels Specifications developed in 2006 were made in line with European, United States, and Japanese standards, as manufacturers from these markets dominate the automotive industry.

A gradual shift of policies: Toward volumetric mandates and subsidies

The late 2000s saw a widespread shift away from tax exemptions to consumption mandates or volumetric subsidies (Bahar et al., 2013). Many countries that had initially implemented tax incentives gradually decreased or removed them because they could lead to important government revenue losses (this was particularly the case in the United States and Korea) and proved to be an insufficient incentive in several economies. Table 2.1 summarizes the levels of biofuel mandates in key countries as of 2018.

Since 2009, the major biofuel policy at the EU level is the 10% blending mandate introduced through the Renewable Energy Directive (RED). In several member states

Table 2.1 Current blending mandates in road transport fuels (as of March 2018)

Country	Reference	Ethanol	Biodiesel
United States	Volume	73 billion liters, of which 16.2 are advanced biofuels	
Brazil	Volume	27%	7%
European Union[a]	Energy content	10% by 2020, of which a maximum of 7% is food based	
Thailand	Volume	–	7%
China	Volume	10%[b]	–
Argentina	Volume	12%	10%
India	Volume	10%	–
Canada	Volume	5%	2%[c]
Indonesia	Volume	2%	20%
South Korea	Volume	–	2,5%
Japan	Volume	500 million liters crude oil equivalent by 2017	–
Australia	Volume	New South Wales: 7%	NSW: 2%

[a]Several EU member countries have also implemented their own blending mandates, which correspond to targets on energy content, volume, or greenhouse gas emissions savings.
[b]The E10 mandate fully covers six provinces and 27 cities in five other provinces.
[c]Biodiesel and hydrogenated vegetable oil.
Source: OECD-FAO Agricultural Outlook database.

the shift had actually occurred before the implementation of this obligation. In Germany, tax exemptions for all biofuels were replaced after only 1 year by a system based on use mandates. Mandatory blending targets referred to as "quotas" were set up to 2015, and biodiesel was subjected to a gradual tax increase—to reach the same level as the diesel energy tax by 2013. The Netherlands introduced a tax-reduction scheme in 2006, for that year only, and since 2007 has supported biofuels primarily by mandatory blending targets. In Sweden, support policies still consist mainly of tax reductions.

In Canada, Korea, and the United States, former tax credits were eliminated between 2008 and 2011 and replaced with consumption mandates and/or production subsidies. In the United States, however, tax credits remained for cellulosic ethanol and biodiesel until 2016. In 2005, Congress established the first Renewable Fuel Standard (RFS) through which the Environmental Protection Agency (EPA) mandates the volumes of biofuels to be supplied annually. In 2007, the enactment of the Energy Independence and Security Act amended the RFS1 and created the RFS2, which remains today the main policy driving the US biofuels sector. Volume requirements were set up to 2022 for four categories of biofuels: *cellulosic biofuel*, *biodiesel*, *total advanced*, and *total renewable fuel*, each subjected to lifecycle GHG emissions thresholds.

In Thailand, an important shift of types of biodiesel policies occurred as mandatory blending replaced price and tax subsidies. The South African government had planned a similar shift in its 2007 Biofuels Strategy (Republic of South Africa, 2007), in which tax exemptions were meant to gradually set the path for a future implementation of blending mandates, which finally took place in 2013. In Australia, a gradual increase in excise duties was also set to operate from 2015 to 2020, but only two states currently have blending mandates.

During the 2000s, a great increase in investments in feedstock cultivation and transformation plants occurred in both developed and developing countries around the world. However, after the kick-off phase and as the industry started to expand, production slowed down in the late 2000s and policies started to shift as several controversies and concerns related to the impacts of biofuels on food markets and land-use changes were brought to the international stage.

2.3 2010–2018

In the late 2000s, two major controversies brought to light potential negative impacts of biofuels. As the 2008 global food crisis unfolded, criticisms emerged at the international scale arguing that biofuel support policies contributed to food price spikes. Because biofuels production is largely policy driven, demand is relatively inelastic to price variations, which also increases the price volatility of agricultural commodities. As the production of biofuels was perceived as a threat to food security, the food versus fuel debate gained momentum in several countries. In 2006, the Chinese government issued an urgent notice that effectively halted the construction of new corn-based ethanol plants. The issue was particularly strong in Japan, which has a low food self-sufficiency rate and depends largely on imports. At the G20 meeting of 2011, joint international organizations issued a recommendation advocating for the reduction of support policies for biofuels and urging to at least introduce flexibility in biofuel mandates and subsidies and diminish trade restrictions (FAO et al., 2011).

The sustainability agenda

Simultaneously, sustainability concerns were arising, focused on the increasing pressure on finite resources such as land and water, and most importantly on land-use changes originating from increasing biofuels production. Indirect land-use changes (ILUC) occur when the conversion of existing crops to cultivation of biofuel feedstocks leads to the expansion of cropland on natural land elsewhere for food production. While initially biofuels had largely been promoted as a solution to reduce GHG emissions from the transport sector, ILUC considerations led to a greater controversy on the actual contribution of biofuels to the mitigation of emissions. Multiple studies were launched around 2009 to examine the extent and consequences of ILUC, notably in Europe and the United States.

Environmental concerns and uncertainties led to a general shift in the focus of policy development, most pronounced in developed countries. In the European Union, concerns about ILUC-related emissions initiated discussions on a cap for conventional biofuels. In 2012, the European Commission advocated for an upper limit of 5%, but the European Council reached a final decision in 2015 and implemented a 7% cap on food-based biofuels. To reach the RED 10% target by 2020, the remaining 3% could come from a variety of alternatives, including advanced biofuels but also renewable electricity in rail and electric vehicles. Policy responses were also strong in several member states: in the United Kingdom, concerns on ILUC impacts led to the downward revision of recent ambitious policies as a precautionary measure: blending obligations under the Renewable Transport Fuel Obligations were substantially decreased in 2009, and tax exemptions introduced in 2002 and 2005 ceased the following year. In the United States, the EPA issued its final rulemaking for 2014 and 2015 with a delay after considering revising RFS mandates downward.

An increasing number of developed countries defined obligations for fuel refiners to supply biofuels that meet certain sustainability criteria with a focus on GHG emissions. New policies based specifically on sustainability criteria were also implemented. In British Columbia, Canada, increasingly strict carbon intensity limits are prescribed annually for transportation fuels. In 2015, in Germany, the biofuel blending mandate has been replaced with minimum GHG savings obligations. In the state of California an original and complex biofuel policy combining three mechanisms—the federal RFS and the state Low Carbon Fuel Standard (LCFS) and Cap and Trade (CAT) program—has been in effect since 2011. In 2009, the California Air Resources Board commissioned a study on the carbon intensity of biofuels—in particular on ILUC-related emissions—and implemented as a result the LCFS, which is a market-based cap and trade approach to lower GHG emissions from transportation fuels. The multisectoral CAT program complements the RFS and LCFS by penalizing fossil fuels; it sets a total emissions allowance budget and lets the market determine where those emissions occur based on a trading mechanism.

A global trend has also recently emerged: policies have contributed to a global reduction in trade since 2012 (Fig. 2.1). In the past decade, trade flows were mostly from developing countries to developed countries. Because the latter are restraining their consumption of first-generation biofuels, their import demand is likely to decrease or stagnate. Additionally, the implementation of sustainability criteria and the introduction of countervailing duties imply that only certified biofuels can be exported to these markets.

Brazil is with the United States a major ethanol exporter. Studies from both the United States and the Japanese governments concluded that Brazilian sugar cane ethanol induced sufficient levels of GHG emissions reduction. Under influence from the industry, domestic Brazilian blending mandates also increased. Other important exporters, such as Argentina, have not implemented official environmental or social sustainability criteria for biofuels; however, governments closely monitor other countries' criteria and regulations to avoid export restrictions. Argentinean and Indonesian biodiesel exports to the European Union and the United States have been or are still the subject of antidumping duties.

In developing countries, where biofuels were initially seen as a means for rural development and economic growth, concerns regarding GHG emissions are also brought forward as an objective of biofuel policies. As a result, production is growing and remains destined for domestic consumption, and because few sustainability restrictions have been implemented, local mandates are increasing. In India both ethanol imports and exports are

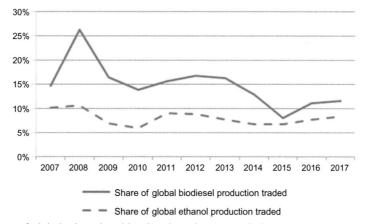

Fig. 2.1 Shares of global ethanol and biodiesel production traded, 2007–17.

discouraged, and in 2015 the 5% blending mandate was raised to 10%. Mandates have been increasing and supportive policies strengthened in several countries, although actual blending levels remain considerably below mandates due to insufficient support for the industry, notably in Argentina and India. In the particular case of Indonesia, whose biodiesel production was founded on both the country's strong position as a palm oil producer and growing European demand for biodiesel (80% of production was shipped overseas in 2009), the importance of exports has diminished mainly because of energy security concerns. Following rises in fossil fuel prices, the government started encouraging domestic biodiesel consumption and set a first mandate in 2006. The new subsidization scheme implemented in 2015 further reduced exports to the benefit of domestic consumption, with the introduction of a biofuel "plantation fund" supplied by a levy on palm oil exports.

Several countries have also shifted to more flexible instruments, as recommended by international organizations at the 2011 G20 summit (FAO et al., 2011). While export tariffs were first set at fixed levels, in 2012 Argentina introduced a flexible export tax system in which the level of the export duty is fixed on a monthly basis, thereby enabling the government to take into account the evolution of international prices and the costs of local production. Other countries adopted flexible mandates, with levels either changing frequently through the usual regulatory process or set to vary within a range. This was already the case in Thailand, whose government has been adjusting the mandate since 2007, depending on the domestic supply and stocks of palm oil, which vary with the season and harvest yield. The state of mandates and policies thus continues to fluctuate, the current mandate has been on and off, and lesser blends—down to "B3.5": 3.5% of incorporation in the biodiesel pool—have been the fallback. In India, sugar industries have been lobbying for a flexible and higher blending level in case of years of high production.

The policy developments lying ahead

The recent (2015) context of low international crude oil prices exacerbated the very limited visibility on biofuel policies. Currently, biofuel use is driven by country-specific biofuel-use mandates rather by market demand. While uncertainties arose on biofuels' potential to contribute to the reduction of GHG emissions, global mitigation goals were increasing as well. Across the world, the focus of policy development in the transport sector looks at other ways to reduce carbon emissions. Broader energy and environmental strategies are now being adopted, focusing notably on the development of electric vehicles and public transportation.

The uncertainties related to biofuels' potential to contribute to the reduction of GHG emissions are important and only a few countries have stated biofuels development in their national commitments to reduce GHG emissions. In particular, the RenovaBio program was officially signed in January 2018 by the Brazilian government as a follow-up to its commitment to reduce GHG emissions by 37% in 2025 and 43% in 2030 compared to their 2005 level. The program defines a minimum blending target for the volume of fuel ethanol that should reach 30% by 2022 and 40% by 2030.

Despite the uncertainties, biofuel policies are certain to evolve in the near future. On January 17, 2018, the European Parliament proposed to reach 12% renewable energy in transport fuel by 2030. This proposal states that the consumption of biofuels based on food and feed feedstock cannot increase above the 2017 levels.[4] Palm oil-based biodiesel would be prohibited after 2021 and the share of advanced biofuels, including those produced from waste, should reach 1.5% by 2021 and 10% by 2030.

In Canada, the federal Renewable Fuels Regulations that mandate 5% renewable content in gasoline and 2% in diesel fuel could be replaced in 2019 by the Clean Fuels Standard. This new regulatory framework, which was presented in December 2017, would use a lifecycle approach to set carbon intensity requirements and would apply to liquid, gaseous, and solid fuels combusted for energy generation, in addition to the transportation sector. The objective is to achieve 30 Mt of annual reductions in GHG emissions by 2030. In September 2017, the Chinese government proposed a nationwide ethanol mandate that would expand the mandatory use of E10 fuel from 11 trial provinces to the entire country by 2020. The underlying rationale for that announcement has not been clearly stated but could be related to abundant grains stocks and to environmental concerns. If fully implemented, these policies could have important impacts on biofuels and agricultural markets.

3. Technology: Overview of biofuel pathways

Significantly different definitions and classifications of biofuels are used by various institutions and governments. These differences are not trivial. They not only reflect the targets set in a given national biofuel policy, but are also the basis for certifications in international trade.

Identical products are thus considered differently among some of the major players in the biofuels market. No classification scheme is universally accepted. One of the most troubling examples might be that of biodiesel in European and American legislations. While the European Union defines advanced biofuels as biofuels "which provide high GHG emission savings with a low risk of causing ILUC, and do not compete directly for agricultural land for food and feed markets" (Directive (EU) 2015/1513, n.d.)— thereby excluding all vegetable oil-based biodiesel—the US RFS includes all biomass-based diesel in its advanced category (Fig. 2.2).

Fig. 2.2 also shows that the RFS refers to sustainability criteria. According to the reference values for lifecycle GHG reductions, the standard classifies sugar cane ethanol as a type of "advanced biofuel."

---

[4] Except for countries whose share of food and feed-based renewables in transportation fuels is below 2%.

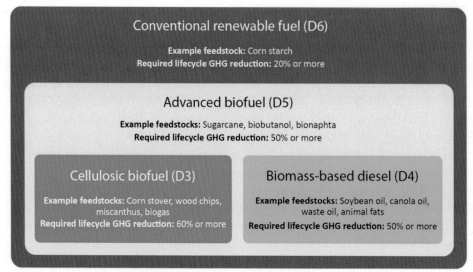

Fig. 2.2 Fuel nesting scheme for the Renewable Fuel Standard. *(Source: US EPA)*

These two products correspond to considerable volumes of biofuels used in the US fleet. The United States consumed about 6,875 million liters of biodiesel in 2015, representing 21.8% of global volumes, mostly derived from soybean oil. Its consumption of ethanol for fuel use averaged 59,230 million liters (representing around 60% of global consumption), of which more than 95% corresponded to sugar cane ethanol, imported mostly from Brazil (USEIA, 2016).

The EU 2015/1513 Directive also makes another distinction among biofuels: fuels derived from used cooking oil and animal fats are deemed "sustainable" (which implies that they can be double-counted toward the general blending mandate) but cannot be counted toward the 0.5% target for "advanced biofuels." In China, where the industry initially developed as a response to an abundance of grain production, a differentiation is now made between grain-based and nongrain-based ethanol.

Furthermore, a wide range of terms are used to refer to biofuels: first-, second-, third-generation, sustainable, renewable, advanced, etc., while definitions in the literature are often inconsistent. These classifications are alternately based on the type of feedstock, the conversion technology used, or the properties of the fuel molecules produced.

This is symptomatic of the immense diversity of feedstocks from which biofuels can be produced, and of the profound changes the sector is undergoing. In addition, the conversion routes overlap; the different transformation processes are not parallel but crossed (Fig. 2.3). For example, renewable diesel (often erroneously referred to as biodiesel) can be produced both from vegetable oil and cellulosic material, while the latter can also be transformed into ethanol, biogas, jet fuel, and even gasoline. A wheat crop could—if technologies were commercialized—lead to the production of both ethanol via

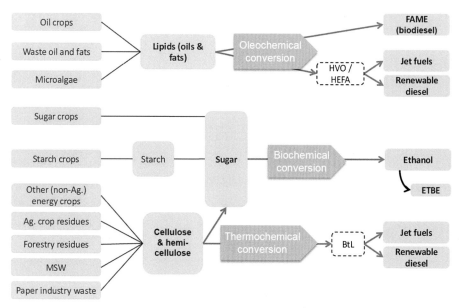

Fig. 2.3 Overview of the main biofuel pathways.

biochemical conversion of the starch present in the grain, and renewable diesel via gasification and Fischer–Tropsch synthesis of the fiber from the straw.

However, government support and economic instruments occur at different steps along the biofuels value chain. From crop subsidies to carbon taxes and on to tax credits for flex-fuel vehicles, the resulting economic and environmental impacts from distinct economic instruments are vastly different. For these reasons, understanding properly the structure of the biofuels value chain is critical. Moreover, while the industry initially developed to respond to various objectives, today biofuel policies must be aligned with a long-term strategy, designed with consideration of climate change mitigation and objectives of an ecological and energy transition. In doing so, three characteristics of biofuels are of interest:

- Which feedstock do they originate from? (This will have subsequent impacts on land use.)
- Which conversion technology is used and what are its production costs? (Reducing these costs is critical to the competitiveness and commercialization of advanced biofuels.)
- What is the end-product? (This will determine which fossil fuel is substituted.)

Figs. 2.3–2.5 present a detailed mapping of existing conversion routes. From these, we define the notion of "biofuel pathways": the combination of three elements—a source of biomass, a conversion process, and an end-product. Fig. 2.3 provides an overview of the three main conversion routes for liquid biofuels: oleochemical conversion, biochemical

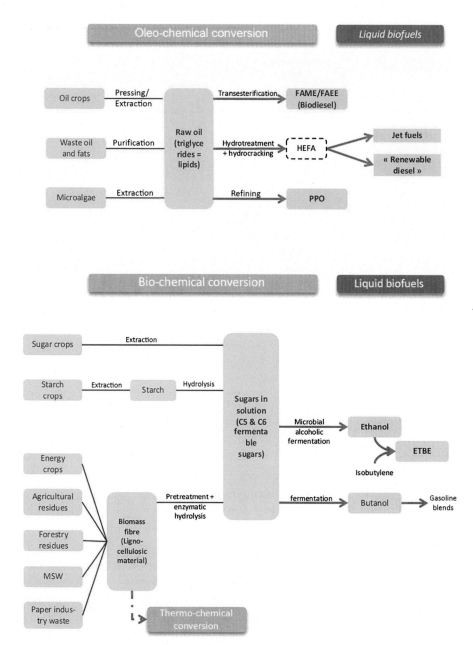

Fig. 2.4 Mapping of the main biofuel pathways: (1) oleochemical and biochemical conversions.

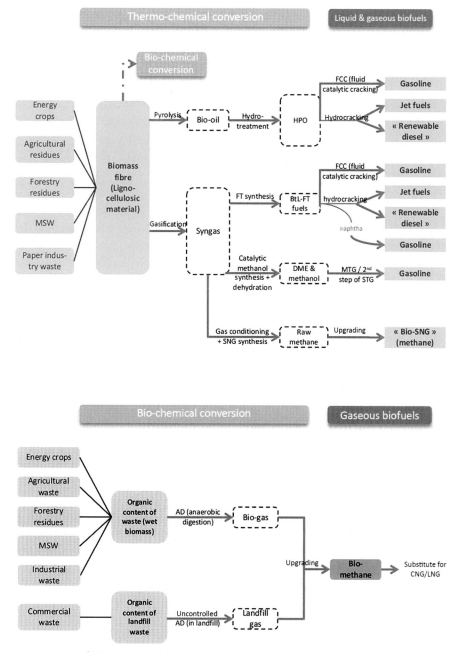

Fig. 2.5 Mapping of the main biofuel pathways: (2) thermochemical conversion (and biochemical conversion of biomethane).

conversion, and thermochemical conversion. Figs. 2.4 and 2.5 provide a more detailed description of these different pathways.[5]

4. Conclusion

This chapter provided an overview of the evaluation of biofuel policies, institutional options, and biofuel production pathways. Developments on biofuel markets have been and are likely to continue to be strongly related to biofuel policy packages in place, the global macroeconomic environment, and the level of crude oil prices. Over the past few years, the biofuel policy environment was relatively uncertain. Recent policy announcements appear to be favorable to biofuels with a focus on sustainability criteria or on the potential contribution of renewable fuels to GHG mitigation in the transport sector. It is not yet clear whether those announcements will imply stronger investments in research and development for advanced biofuels produced from lignocellulosic biomass, waste, or nonfood feedstock. These cannot currently compete with established biofuels due to high production costs. At present, most biofuels produced are still based on agricultural feedstock with direct and indirect effects on the environment, land use, and to a certain extent on agricultural markets. Increasing the use of nonfood-based feedstocks that meet the required sustainability criteria will require a stronger institutional framework.

Disclaimer

The views expressed are those of the author and do not necessarily represent the official views of the OECD or of the governments of its member countries.

References

Bahar, H., Egeland, J., Steenblik, R., 2013. Domestic incentive measures for renewable energy with possible trade implications. In: OECD Trade and Environment Working Papers.

Directive (EU) 2015/1513 of the European Parliament and of the Council of 9 September 2015 amending Directive 98/7/EC relating to the quality of petrol and diesel fuels and amending Directive 2009/28/EC on the promotion of the use of energy from renewable sources (Text with EEA relevance).

FAO, IFAD, IMF, OECD, UNCTAD, WFP, the World Bank, the WTO, IFPRI, the UN HLTF, 2011. Price Volatility in Food and Agricultural Markets: Policy Responses.

International Energy Agency, 2016. Energy Technology Perspectives report 2016: Towards Sustainable Urban Energy Systems. OECD.

OECD-FAO (2017), OECD-FAO Agricultural Outlook Database accessed online http://stats.oecd.org/Index.aspx?datasetcode=HIGH_AGLINK_2017, March 2018.

Republic of South Africa, Department of Minerals and Energy, 2007. Biofuels Industrial Strategy of the Republic of South Africa.

---

[5] Through the conversion process, the product undergoes three main stages: pretreatment, conversion, and refining. Only the main phases of the conversion and refining processes are described here. Additional steps such as extraction, posttreatment, cleaning, etc. occur as well.

United States Energy Information Administration. 2016 Petroleum & Other Liquids—U.S. Imports by Country of Origin—Fuel Ethanol (Renewable). https://www.eia.gov/petroleum/data.php. Accessed 1 August 2016.

Further reading

United States Environmental Protection Agency, Renewable Fuel Standard Program—Renewable Fuel Annual Standards. https://www.epa.gov/renewable-fuel-standard-program/renewable-fuel-annual-standards. (Accessed 1 August 2016).

SECTION 2

Technological options

CHAPTER 3

From biomass to biofuel economics

Deepayan Debnath
Food and Agricultural Policy Research Institute, University of Missouri, Columbia, MO, United States

Contents

1. Introduction

Developing technological, institutional, and policy interventions to balance the food and bioenergy contribution of food systems requires a full understanding of the process with which biomass is converted into biofuels. Such processes identify the stages in the value chain for the biofuels and the opportunities for interventions based on national and sectoral goals of the country concerned. In this chapter, we look at these processes from a technological perspective and identify the opportunities for interventions to make the food system more efficient and sustainable.

So far biofuels are broadly categorized into three categories: (1) first-generation, (2) second-generation, and (3) third-generation biofuels. These classifications are based on the biomass or feedstock uses in the production of biofuel. Even though the production and consumption of biofuel are heavily driven by policies, the profit-maximizing producers are always looking for the product that produces the highest returns, and the biofuel industry is no exception. In the production of biofuels, producers are more likely to use inputs that are easily available in the markets rather than use new feedstocks that either require different production capacities or need to rely on a few specific feedstock suppliers. This leads to the fact that until now the biofuels ethanol and biodiesel available at gas stations across Europe, the United States, and Brazil are first-generation biofuels that are derived from readily available food crop-based feedstocks such as maize, sugar cane, soybeans, and vegetable oils.

Conventionally, any particular country chooses one particular crop for the production of a certain type of biofuel depending on many factors. Quantity produced is one of the factors that drive the choice of the feedstock for that country. In the case of the United States, the majority of the ethanol blended in gasoline is derived from maize because the United States ranks as the major maize producer (FAOSTAT, 2013). In Brazil, sugar cane is used to produce ethanol and the country is also the largest producer of sugar cane. More generally, Brazil has a comparative advantage in producing certain other food crops that can be used for biodiesel, this is because of its geographical location and availability of resources, including land, labor, and capital.

The use of conventional ethanol or biodiesel is limited to a few countries. In this chapter, first-generation biofuel is referred as conventional ethanol and now onwards this term will be used. Conventional starch-based ethanol derived from grains and sugar cane is used in European Union countries, the United States, and Brazil, while biodiesel is mainly used in the European Union and the United States. In these three countries, conventional ethanol use is dictated by governmental policy. Recent economic debates have fueled the question that, in the absence of policy, is conventional ethanol able to compete with unleaded gasoline at the pump (retail level)? To answer this question, many aspects of biofuel industry need to be discussed, including the conversion technology discussed in this chapter. The scope of this chapter is limited to the discussion of biofuel in the context of liquid fuel, which can be used for motor vehicles.

This chapter is mainly focused on the economics of the use of food crops for the production of biofuel. The conversion technologies used to convert biomass (maize, sugar cane, soybeans, vegetable oils, etc.) to biofuel (ethanol and biodiesel) and their economic and environmental implications is discussed.

The reader should get an idea of the economics of the production of first-generation biofuel and should be aware of its advantages and disadvantages.

2. Conversion of maize to ethanol: A closer look

By definition, alcohol produced from sugar by yeast is ethanol. Conventional ethanol can be produced from any plants that contain sugar. Ethanol has all the identical chemical and physical properties of other alcoholic beverages such as beer, wine, and spirits. The only difference between fuel ethanol and alcohol is that the water content is removed from the fuel ethanol and it is denatured by blending either with benzene, olefins, or aromatics to make it unfit for human consumption (Renewable Fuel Association, 2010). Following the process of biochemical conversion, the plant sugar can be fermented into ethanol. However, the problem lies in making fuel ethanol economically efficient.

An important microorganism to humans is the yeast *Saccharomyces cerevisiae*, which is actively used for the process of the fermentation of maize to produce ethanol. Through this process, the yeast simply breaks the sugar content in the maize into carbon dioxide (CO_2) and ethanol (C_2H_6O). One kilogram of sugar in maize, through the process of fermentation, produces 0.5 kg of ethanol and the same amount of CO_2, that is, ethanol yield from sugar is 50% of the actual volume. Based on this conversion rate, 1 kg of maize produces 0.42 L of ethanol, and depending on milling technology it generates another coproduct called "distiller's dried grains with solubles (DDGS)". We will discuss DDGS in the next section.

In many countries, including the United States, fuel ethanol is commercially produced by breaking down the starch present in maize into simple sugars (glucose); these sugars are then mixed with yeast (fermentation), and finally the main product bioethanol and by-products such as DDGS (animal feed) are recovered. There are two major industrial methods for producing fuel ethanol from maize: (1) wet milling and (2) dry grind.

The United States, which is the leading maize-based bioethanol-producing country, uses dry-grind ethanol production processing technology. More than 70% of fuel ethanol produced in the United States is from dry-grind processing plants. The main reason is that dry grind biorefineries require less investment and can be built on a much smaller scale compared to the wet milling processing plant.

2.1 Wet milling

Besides producing fuel ethanol, wet milling is used to produce many products. Compared to dry grind, maize-processing wet mills are large scale and capital intensive, and produce varied products such as high fructose maize syrup, biodegradable plastics, food additives such as citric acid and xanthan gum, maize oil used for cooking oil, and livestock feed too.

The first step in the wet milling process involves soaking the grain in water (steeping). For that reason, the word "wet" is used for this milling process. Soaking softens the grain and makes it easier to separate the various components of the maize kernel. The process of separating soaked maize into many other components is called "fractionation." The process of fractionation separates wet maize grains into starch, fiber, and germ. Once they are separated, they are processed to make a variety of products. The major by-products of wet milling fuel ethanol production are two commercially available feed products: maize gluten meal, which contains high protein (40%) and is used for human consumption, maize gluten which contains low protein (28%) and is used as animal feed; and maize germ, which is further used to produce maize oil used for cooking purposes.

2.2 Dry grind

In the dry-grind ethanol process, the whole grain is processed, and the residual components are separated at the end of the process. There are six major steps in the dry-grind method of ethanol production:
1. Milling and liquefaction.
2. Saccharification.
3. Fermentation.
4. Distillation.
5. Rectification.
6. Dehydration.

Milling and liquefaction

Milling involves processing maize through a hammer mill with screens between 3.2 and 4.0 mm to produce a flour (Rausch et al., 2005). This maize flour is then slurried with water and heat stabilized along with an enzyme called α-amylase. The slurry is then cooked through a process known as "liquefaction." It is performed using jet cookers that inject steam into the flour slurry to cook it at temperatures above 100°C. The heat and mechanical shear of the cooking process breaks apart the starch granules present in the kernel endosperm, and the enzymes break down the starch polymer into small fragments. The cooked maize mash is then allowed to cool to 80–90°C, the additional α-amylase enzyme is added, and the slurry is allowed to continue liquefying for at least 30 min.

Saccharification

After milling and liquefaction, the maize slurry is now called "mash," which is furthered cooled to approximately 30°C, and a second enzyme, glucoamylase, is added to the mash. Glucoamylase completes the breakdown of the starch into a simple sugar (glucose). This step is called "saccharification" and often occurs while the mash is filling the fermenter in preparation for the next step of fermentation and continues throughout the next step.

Fermentation

In this step, yeast grown in seed tanks is added to the previously processed mash to begin the process of converting the simple sugars to ethanol. The other components of the kernel (protein, oil, etc.) remain largely unchanged during the fermentation process. In most dry-grind ethanol plants, the fermentation process occurs in batches. A fermentation tank is filled, and the batch ferments completely before the tank is drained and refilled with a

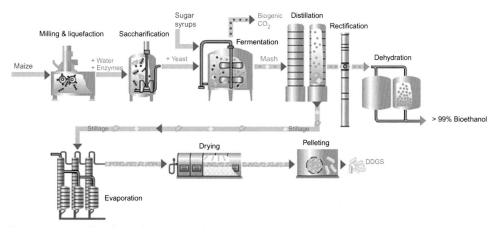

Fig. 3.1 Dry-grind milling of maize to ethanol.

new batch. The upstream processes of milling and liquefaction, saccharification and downstream process distillation, and rectification and dehydration occur continuously through the equipment (as shown in Fig. 3.1). Thus dry-grind facilities of this design usually have three fermenters (tanks for fermentation) where at any given time one is filling, one is fermenting, which usually takes 48 h, and the last one is emptying and resetting for the next batch. Carbon dioxide is also produced during fermentation. Usually, the carbon dioxide is not recovered and is released from the fermenters to the atmosphere. If recovered, this carbon dioxide can be compressed and sold for the carbonation of soft drinks or frozen into dry ice for cold product storage and transportation. After the fermentation is complete, the fermented corn mash more popularly known as "beer" is emptied from the fermenter into a beer well. The beer well stores the fermented beer between batches and supplies a continuous stream of material to the ethanol recovery steps including distillation.

Distillation

After fermentation, the liquid portion of the slurry has 8%–12% ethanol by weight. Because ethanol boils at a lower temperature than water, the ethanol can be separated by a process called "distillation." Conventional distillation systems can produce ethanol at 92%–95% purity.

Rectification

The azeotropic composition of an ethanol/water mixture obtained from distillation at atmospheric pressure contains about 95% of ethanol, which is the maximum achievable purity of ethanol in an atmospheric rectification column. The residual water is then further removed using a molecular sieve process called "rectification" that selectively the water from an ethanol/water vapor mixture, resulting in nearly pure ethanol (>99%).

Dehydration

Pure ethanol can not be obtained from distillation since it forms an azeotrope with water at 95%. Fuel ethanol or absolute alcohol is produced by dehydration of rectified spirit (ethanol). Commercially available technologies for the dehydration of rectified spirit are azeotropic distillation and molecular sieve technology:

Azeotropic distillation: To dehydrate ethanol from azeotropic concentration, a third substance called an entrainer (trichloroethylene, benzene, toluene, cyclohexane, etc.) is added to the mixture of ethanol and water. An entrainer breaks the azeotropic point of ethanol and water, i.e., it alters the relative volatility of water making it more volatile. The ternary azeotropic mixture formed at the top of the dehydration column, allows the removal of water and thus dehydrates the alcohol. The azeotropic mixture is heterogeneous and the "heavy" phase which is high in water content is extracted by decantation. The regeneration column allows water extraction from the "heavy" phase and entrainer recycling.

Molecular sieve technology: Molecular sieve technology works on the principle of pressure swing adsorption. Here water is removed by adsorption on the surface of "molecular sieves" under pressure and then cyclically removed under low pressure at different conditions. This process carries out dehydration of mixed ethanol and water by adsorption of water into zeolite balls, which are molecular sieves. The dehydration unit operates with two adsorbers according to alternate steps of adsorption and desorption. Adsorption occurs in the vapor phase and under pressure. Desorption regenerates water-saturated molecular sieves. This step is performed under vacuum. Part of the dehydrated alcohol is used for molecular sieve desorption. Alcoholic effluent from desorption is regenerated within the distillation column.

The residual water and solid maize that remain after the distillation process are called "stillage." The stillage is then centrifuged to separate the liquid (thin stillage) from the solid fragments of the kernel (wet cake or distillers grains). Some of the thin stillage (backset) is recycled to the beginning of the dry-grind process to conserve the water used by the facility. The remaining thin stillage passes through evaporators to remove a significant portion of the water to produce thickened syrup. This process is often called "evaporation." Usually, the syrup is blended with the distillers grains and dried through the process of "dying" to produce DDGS. When feedlots are close to the plant, the by-product may be sold without drying as distillers grains or wet distillers grains.

3. Conversion of sugar cane to ethanol

The conversion of sugar cane to ethanol consists of two steps: preprocessing and processing.

Preprocessing: The first step is harvesting cane. In many sugar-producing countries, including Brazil and India, mechanical harvesters are used to cut the cane into small pieces and load it into trucks during the harvesting operation. Trucks then deliver the harvested sugar cane to the processing mill. Each truck is weighed to check how much cane it is delivering. Samples are removed to test the quality of the cane. Sugar cane must be processed as soon as possible to avoid its sugar content to deteriorate. Most cane is delivered within 24 hours of harvesting. One ton of sugar cane produces around 85 L of ethanol. The truck unloads the cane onto preparation conveyor belts that carry the cane to the crushing system. Sugar cane juice is extracted by crushing the chopped cane through a series of rollers and the juice flows out. Cane fiber called bagasse goes to boilers to be burned. The heat turns water into high-pressure steam for electricity. Bagasse in the ethanol industry is used for electricity production.

Processing: In this stage the sugar cane juice is heated and treated with sulfur, lime, and thickener. The mixture is pumped to rotating filters that separate the juice from most impurities. These form a crumbly residue, known as filter cake, which is used as natural fertilizer on fields. The filtered juice passes through sieves that remove any remaining impurities. The pure juice then goes to the evaporation process to remove the extra water content. Finally, the thick sugar juice goes either to make sugar or to make ethanol. This process is very popular in Brazilian sugar mills where both sugar-processing and ethanol-fermenting facilities exist under one roof. Therefore the mill managers make the decision as to how much sugar and ethanol to produce depending on the demand for each (sugar and ethanol) product. Water and yeast are added to ferment the liquid sugar cane juice to produce ethanol. Like maize starch-based ethanol the fermented liquid, called "beer," passes through centrifuges. The yeast is removed, treated, and reused. The liquid flows through two distillation columns that heat it to remove water. This produces dehydrated ethanol, which is blended with petrol for use as a transport fuel. The liquid by-product, called vinasse, is sprayed onto fields as fertilizer. At the pump, Brazilian motorists choose between pure ethanol and a petrol/ethanol blend. Some liquid flows into a third distillation column that further heats the liquid to remove more water. This produces anhydrous ethanol, which contains less water. It is considered to be >99% pure bioethanol and could be blended with gasoline at any percentage. The efficient industrial mills process sugar cane into ethanol in around 12–15 hours. The ethanol flows into tanks to be stored and then transported to the market. If the mill manager chooses to produce sugar, then the sugar cane juice is further cooked to evaporate the last portion of the water under very tight controls in a vacuum pan. Seed grain (pulverized sugar) is fed into the pan as the water evaporates and crystals begin to form. The mixture leaves the vacuum pan as a thick crystal mass and is sent to a centrifuge, a large perforated basket spinning very rapidly much like a washing machine in the spin cycle, where it is spun and dried, yielding golden raw sugar, which is bagged according to weight. The entire conversion process is depicted in Fig. 3.2.

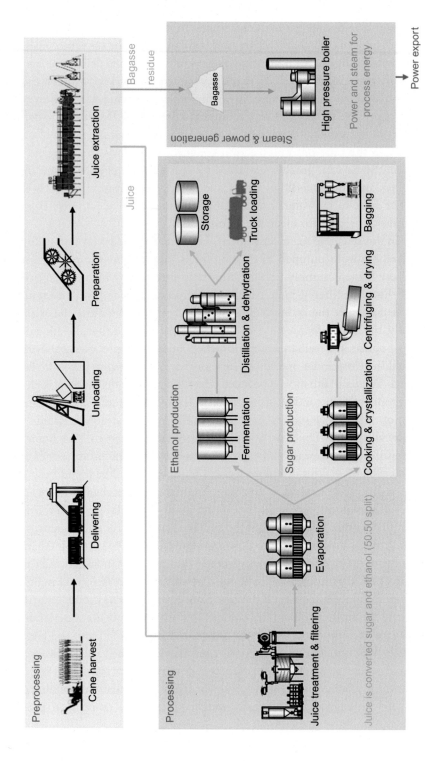

Fig. 3.2 Conversion of sugar cane to ethanol and sugar.

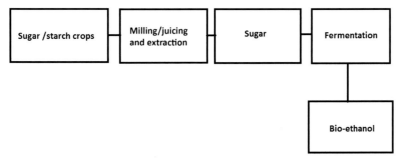

Fig. 3.3 Schematic diagram of the conversion of grains to ethanol.

4. Conversion of other grains to ethanol

Other grains, mainly wheat and barley, are used extensively in Canada and many parts of Europe to produce ethanol. The process of harvesting gains varies according to the grain. Once the sugar is extracted from the grains the ethanol is obtained through the process of fermentation. The more general structure of converting grains to ethanol is shown in Fig. 3.3, which shows that any crops that contain a significant quantity of sugar starch could go through the process of extraction and fermentation to generate ethanol.

5. Biodiesel production system

Biodiesel has a viscosity similar to petroleum diesel and can be used as an additive in formulations of diesel to increase lubricity. Biodiesel can be used in pure form (B100) or may be blended with petroleum diesel at any concentration in most modern diesel engines. Biodiesel will degrade natural rubber gaskets and hoses in vehicles (mostly found in vehicles manufactured before 1992), although these tend to wear out naturally and most likely will have already been replaced with Viton-type seals and hoses, which are nonreactive to biodiesel. Biodiesel's higher lubricity index compared to petroleum diesel is an advantage and can contribute to longer fuel injector life.

Biodiesel is a better solvent than petroleum diesel and has been known to break down deposits of residue in the fuel lines of vehicles that have previously been run on petroleum diesel. Fuel filters may become clogged with particulates if a quick transition to pure biodiesel is made, because biodiesel "cleans" the engine in the process. It is therefore recommended to change the fuel filter within 600–800 miles after first switching to a biodiesel blend.

Biodiesel's commercial fuel quality is measured by the ASTM standard designated D 6751. The standards ensure that biodiesel is pure and the following important factors in the fuel production process are satisfied:
• Complete reaction
• Removal of glycerin

- Removal of catalyst
- Removal of alcohol
- Absence of free fatty acids (FFAs)
- Low sulfur content

Biodiesel is at present the most attractive market alternative among the nonfood applications of vegetable oils for transportation fuels. The different stages in the production of plant/seed oil methyl ester generate by-products that offer further outlets. Oil cake, the protein-rich fraction obtained after the oil has been extracted from the seed, is used for animal feed. Glycerol, the other important by-product, has numerous applications in the oil and chemical industries such as the cosmetic, pharmaceutical, food, and painting industries.

Oil crops are the base for biodiesel production. In Europe, rapeseed is the most common feedstock for biodiesel production. In the United States, Argentina, and Brazil, soybean oil is the most dominant biodiesel fuel feedstock. In Indonesia and Malaysia, palm oil is the main feedstock cultivated. Besides the most prominent oil crops (palm, soybean, rapeseed, and sunflower), many other crops such as canola, mustard, flax, jatropha, coconut, hemp, and pennycress are good sources of oil. The use of corn oil is also gaining momentum in the United States, where large volumes of maize are used in ethanol production.

5.1 Conversion of oilseed to biodiesel

Oilseed crops store energy from the sun through photosynthesis during the process of producing their own food. With the help of modern sophisticated technologies, those energy stores in the plant are converted into biodiesel used to run cars. The seeds are harvested from the oilseed crops and the harvesting mechanism differs from one oilseed-producing crop to another. The next step is cleaning, drying, and preparing the seeds for crushing. Foreign particles such as stone, glass, and metal are removed, and then drying is performed while avoiding contact with combustion gases unless natural gas is used. In the solvent extraction process, the clean and dried seeds are crushed and heated to break them into two components: oil and solvent. The process separates the oil from the seeds/beans. Seeds with high oil content, such as rapeseeds and sunflower seeds, are usually mechanically pressed after a preheating step. The pressed cake contains up to 18% of oil and is further treated in the extractor. In some cases, the pressed cake undergoes deep expelling. This brings oil levels down to below 10% and results in an expeller sold for feed purposes. Soybeans have a relatively low oil content. They are thermally treated and mechanically crushed into raw materials/flakes that are further extracted. Sometimes the oil-containing raw material is pressed without heating; such oils are known as cold-pressed oils. Since cold pressing does not extract all the oil, it is practiced only in the production of a few special edible oils for example olive oil.

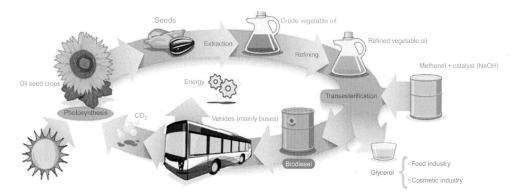

Fig. 3.4 Conversion of oilseed crops into biodiesel.

The preprocessed seeds/beans are treated in a multistage countercurrent process with solvent until the oil content is reduced to the lowest possible level. Hexane is commonly used as extraction solvent. The miscella is a mixture of oil and solvent. It is separated by distillation into its two components: crude vegetable oil and solvent. The solvent is recycled for reuse in the extraction process. The crude oils are then refined to remove the gums or crude lecithins and FFA from the oil to obtain a neutral taste of the edible oil while maintaining the nutritional value and ensuring the quality and stability of the refined vegetable oil. The biodiesel is derived from vegetable oil through the process of transesterification. Along with producing biodiesel the process generates glycerol as a by-product, which is used extensively in both the food and cosmetics industries. The entire procedure of conversion of oilseed crops to biodiesel is shown in Fig. 3.4.

5.2 Transesterification

Almost all biodiesel is produced by the process called "transesterification." It is the most economical process requiring only low temperatures and pressures while producing a 98% conversion yield. The transesterification process is the reaction of a triglyceride (fat/oil) with an alcohol (methanol) to form esters and glycerol. A triglyceride has a glycerin molecule as its base with three long-chain fatty acids attached. The characteristics of the fat are determined by the nature of the fatty acids attached to the glycerin. The nature of the fatty acids can in turn affect the characteristics of the biodiesel. Transesterification of natural glycerides with methanol to methyl esters is a technically important reaction that has been used extensively in the soap and detergent manufacturing industry worldwide for many years.

During the esterification process, the triglyceride is reacted with alcohol in the presence of a catalyst, usually a strong alkaline like sodium hydroxide. The alcohol reacts with the fatty acids to form the monoalkyl ester or biodiesel and crude glycerol. In most production methanol or ethanol is the alcohol used (methanol produces methyl esters, ethanol produces ethyl esters) and is base catalyzed by either potassium or sodium hydroxide. Potassium hydroxide has been found more suitable for ethyl ester biodiesel production, but either base can be used for methyl ester production.

The products of the reaction are the biodiesel and glycerol. A successful transester-ification reaction is signified by the separation of the methyl ester (biodiesel) and glycerol layers after the reaction time. The heavier coproduct glycerol settles out and may be sold as it is or purified for use in other industries, for example pharmaceuticals, cosmetics, and detergents. After the transesterification reaction and the separation of the crude heavy glycerin phase, the producer is left with a crude light biodiesel phase. This crude biodiesel requires some purification prior to use.

6. Economic efficiency of food-based ethanol and biodiesel

The economic efficiency of the use of feedstock maize, sugar cane, wheat, barley, etc for the production of ethanol and vegetable oil as feedstock including soybean oil, palm oil, rapeseed oil, etc for the production of biodiesel, depends on two important things: con-version rates and yield. These two factors are discussed next.

6.1 Ethanol and biodiesel yield

In the process of converting food grains into bioethanol, along with ethanol other by-products including distillers grain contribute a significant portion. The amount of starch or sugar content in the food grains or sugar cane and sugar beet drives the ethanol yield. The average conversion rates of ethanol range from 0.11 (sugar beet-based feed-stock) to 0.40 (maize-based feedstock). Table 3.1 shows the different ethanol conversion rates based on different food grains. One metric ton (MT) of maize can produce 0.40 MT of ethanol, 1 MT of wheat can produce 0.34 MT of ethanol, 1 MT of wheat or rye can produce 0.19 MT of ethanol, 1 MT of sugar cane can produce 0.07 MT of ethanol, and 1 MT of sugar beet which is popularly used in France can produce 0.11 MT of ethanol. Based on Table 3.1, it is impossible to conclude which food crops have the highest bioethanol yield. To come to this conclusion, crop yield needs to be taken into account.

Table 3.1 Crop-specific ethanol conversion rate

Crops (MT)	Ethanol conversion rate (MT)
Maize	0.40
Wheat	0.34
Barley	0.19
Rye	0.19
Sugar cane	0.07
Sugar beet	0.11

Table 3.2 Vegetable oil-specific biodiesel conversion rate

Vegetable oil (MT)	Biodiesel conversion rate (MT)
Soybean oil	0.97
Rapeseed oil	1.00
Palm oil	0.89
Cottonseed oil	1.00

Unlike ethanol where the starch or sugar content of the feedstock drive the conversion rates, in the case of vegetable oil-based biodiesel, the conversion rate is more or less 1:1 even though the process of transesterification does generate by-products in the form of glycerol. Table 3.2 shows different conversion rates of vegetable oil-based feedstock to biodiesel. One MT of soybean oil can be transformed into biodiesel at the rate of 0.97 MT. One MT of rapeseed oil which is a popular oilseed crop in Germany can produce 1 MT of biodiesel. Palm oil is converted to biodiesel at the rate of 0.89 MT. Latin America, including Brazil and Argentina, recently started converting cottonseed oil to biodiesel at the rate of 1:1. Biodiesel yield based on Table 3.2 alone is inconclusive because oil yields vary across the commodity.

6.2 Crops/vegetable oil yields

The calculation of ethanol and biodiesel yield relies on the crops and vegetable oil yields per hectare (ha). Table 3.3 lists the yields of major food crops used to produce ethanol and major vegetable oil yields used to produce biodiesel. Sugar cane has the highest yield of 65 MT/ha making it the most favorable crop to be used for ethanol production in Brazil (the largest cane-producing country). Followed by sugar beet with a yield of 46 MT/ha;

Table 3.3 Crops and vegetable oil yields

Crops/vegetable oil	Yield (MT/ha)
Maize	5.50
Wheat	3.25
Barley	2.82
Rye	2.91
Sugar cane	65.00
Sugar beet	46.00
Vegetable oil	
Soybean oil	0.38
Rapeseed oil	1.00
Palm oil	5.00
Cottonseed oil	0.27

Table 3.4 Biofuel yields in MT/ha

Crops/vegetable oil	Biofuel	Biofuel yield (MT/ha)
Maize	Ethanol	2.20
Wheat	Ethanol	1.11
Barley	Ethanol	0.54
Rye	Ethanol	0.55
Sugar cane	Ethanol	4.55
Sugar beet	Ethanol	5.06
Vegetable oil		
Soybean oil	Biodiesel	0.36
Rapeseed oil	Biodiesel	1.00
Palm oil	Biodiesel	4.45
Cottonseed oil	Biodiesel	0.27

maize, wheat, barley, and rye have a world average yield of 5.5, 3.25, 2.82, and 2.91, respectively. Among vegetable oil yield,[1] palm oil has the highest yield of 5 MT/ha, followed by rapeseed oil at 1 MT/ha, soybean oil 0.38 at MT/ha, and cottonseed oil at 0.27 MT/ha, respectively.

6.3 Food crop-based biofuel yield

Lastly, by multiplying the crops and vegetable oil yields (Table 3.3) with the ethanol and biodiesel conversion factors (Tables 3.1 and 3.2), the biofuel yield can be obtained, as shown in Table 3.4. Ethanol yield ranges from 0.54 MT/ha (rye) to 5.06 MT/ha (sugar beet). Ethanol yield derived from maize, wheat, barley, and sugar cane is 2.20, 1.11, 0.54, and 4.55, respectively. Biodiesel yield ranges from 0.27 MT/ha (cottonseed oil) to 4.45 MT/ha (palm oil). Soybean oil-based feedstock produced 0.36 MT/ha of biodiesel, while rapeseed oil-based feedstock produced 1 MT/ha of biodiesel. Therefore based on these conversion rates, sugar beet is considered to be the most competing feedstock for the production of ethanol, and palm oil-based feedstock the most efficient. However, the use of feedstock is very much restricted due to geographical location. For example, countries like the United States and Canada cannot use palm oil-based feedstock to produce biodiesel even though it is the most efficient, because these countries do not produce enough palm oil to be diverted from its current use to produce biodiesel.

The decision to use specific feedstock for the production of ethanol and biodiesel relies more on the availability of the feedstock rather than economic efficiency. Traditionally, the country that has the comparative advantage of producing a certain crop uses that crop as a feedstock for the production of ethanol and biodiesel, for example the United States being the largest producer of maize uses it as the major feedstock for

---

[1] Vegetable oil yield considering the extraction or crushing rate.

the production of ethanol, where as the largest producer of sugar cane, Brazil uses sugar cane-based feedstock to produce ethanol, and other countries follow a similar path.

7. Advantages and disadvantages of using food-based feedstock

There are advantages and disadvantages of using food crops as a feedstock for the production of both ethanol and biodiesel. This section will shed some light on those advantages and disadvantages.

7.1 Advantages

The advantages of ethanol use for energy production are diverse. The most important is that ethanol contains oxygen and therefore its combustion in engines is more complete. This results in a substantial reduction in carbon monoxide emissions into the atmosphere. Carbon dioxide is released into the atmosphere when ethanol is burned; at least the same amount of carbon dioxide is easily reabsorbed by growing plants during the process of photosynthesis. This completes the natural carbon cycle and helps to reduce the greenhouse effect.

Furthermore, there is the prospect of economic growth whereby a country can focus on diverse energy production to produce ethanol from grains or sugar cane rather than on fossil fuels. The production of feedstock that is used for ethanol production could create employment in the rural sector, as it uses resources that are domestically available.

7.2 Disadvantages

Any policy in a particular country that encourages the use of ethanol or biodiesel, such as the ethanol and biodiesel blend mandates in the United States, the European Union, and Brazil which can be described as a program that stimulates the production of alcohol as an energy source in those countries could lead too dramatic land-use changes. Nevertheless, the area in which grains, vegetable oil, and sugar cane production required as feedstock by the biofuel production facilities has been subjected to the negative effects of a large monoculture crop. A monoculture is the growth of only one species of crop, grown densely over a large land area. As such, monocultures require increased use of pesticides, since the area would be an ideal location for crop pests and diseases to flourish.

Furthermore, it requires vast areas of land and therefore can lead to the destruction of natural habitats. By allowing very fertile agriculture production areas, such as Sao Paulo in Brazil for sugar cane production and the Mid-west in the United States for maize and soybean production, to be devoted to grain or sugar cane would necessarily force other crops out of the area, driving up the price of traditional food crops. Thus not only are traditional food crops forced to move to other areas, the price of land surrounding the feedstock crop plantations has seen a dramatic increase in the creation of biofuel use policy. Moreover, the practice of subsidizing one primary crop for ethanol production, especially a crop that is dominated by a relatively few large-scale farmers, implies the denial of similar subsidies to other crops or producers (Roehr, 2001).

8. Conclusions

This chapter focused on the first-generation biofuels, which are derived from sugar and starch-based crops. Commercially produced ethanol is derived mostly from maize in the United States and from sugar cane in Brazil. Vegetable oils, including soybean oil, rapeseed oil, and palm oil-based feedstocks, are used in the commercial production of biodiesel. The use of feedstock for the production of biofuel differs from country to country and relies on the principle of comparative advantage, rather than on the theory of economic efficiency, that is, a higher yield is better.

Irrespective of the choice of feedstocks, fermentation is the key step in converting crop-based starch and sugar into ethanol, while distillers grain is the major by-product. Biomass-based biodiesel is derived through the process of transesterification. Glycerol is the major by-product derived from biodiesel production. Even though the conversion technology is similar across different categories of feedstock, the procurement process for feedstock varies largely across the crops and the countries.

First-generation biofuels have their pros and cons. The major advantage of using food crop-based feedstock is that there is already an existing market, and procurement of feedstocks is relatively easier compared to the second- and third-generation biofuels. If in the future the demand for nonrenewable fuels declines, then it would not affect feedstocks producers because they could easily divert their crops to the feed and food market. However, critics suggest that the use of food crops for the production of biofuel is a direct threat to global food security and they even think that the cause of the 2008 global food crisis was because of the divergence of food crops to the production of biofuels.

References

FAOSTAT, 2013. Crops Production Data. http://faostat3.fao.org/browse/Q/QC/E.
Rausch, K.D., Belyea, R.L., Ellersieck, M.R., Singh, V., Johnston, D.B., Tumbleson, M.E., 2005. Particle size distributions of ground corn and DDGS from dry grind processing. Trans. ASAE 48 (1), 273–277.
Renewable Fuel Association, 2010. Fuel Ethanol: Industry Guidelines, Specifications, and Procedures, https://ethanolrfa.org/wp-content/uploads/2015/10/Industry-Guidelines-Specifications-and-Procedures.pdf.
Roehr, M., 2001. The Biotechnology of Ethanol: Classical and Future Applications. Wiley-VCH.

CHAPTER 4

Interaction between biofuels and agricultural markets

Deepayan Debnath*, Céline Giner[†]

*Food and Agricultural Policy Research Institute, University of Missouri, Columbia, MO, United States
[†]Trade and Agriculture Directorate, Organisation for Economic Co-operation and Development, Paris, France

Contents

1. Introduction

Biofuels serve as a bridge between the agricultural and energy markets. Biofuel in the form of ethanol was first introduced in the early 19th century. Two important factors may drive biofuel use: (1) biofuels are alternative sources of energy to fossil fuels, (2) it may favor energy independence, and (3) biofuels may emit fewer greenhouse gas (GHG) compared to nonrenewable fuels. The use of biofuels increases with the increase in fossil fuel price. In 1973, when there was a global oil crisis, the Brazilian government introduced ethanol policies to convert sugar into ethanol and use it as a substitute for gasoline in an attempt to protect consumers from soaring oil price but also to help sugar cane producers who were struggling to shed their stocks due to low sugar prices. In 2007, the United States Congress passed the Energy Independence and Security Act, which served as the backbone of the US biofuels policy and tied bioenergy to the agricultural sector. Many scientists blamed biofuels for the 2008 and 2010 spike in the price of grains, which fueled the "Food-versus-fuel debate."

Before the emergence of biofuels, the connection between the energy and agricultural sectors was weak in the absence of direct link between these two sectors. The biofuels sector connects these two markets as agricultural commodities are the major

Biofuels, Bioenergy and Food Security
https://doi.org/10.1016/B978-0-12-803954-0.00004-8

feedstocks in the production of biofuels, while biofuels (ethanol and biodiesel) serve as a complement and/or substitute for fossil fuels.

Biofuels policies and energy prices are the two major sources of volatility in biofuels markets. Domestic country-specific biofuels policies, which dictate the volumetric mandate of ethanol and biodiesel to gasoline and diesel are one of the crucial factors behind the demand for biofuels. Energy price may also contribute to driving biofuels use because high gasoline prices could lead to the use of alternative cheaper biofuels as a substitute, while lower gasoline prices increase gasoline use, which in turn increases complementary biofuels use.

Most of the biofuels produced are based on food feedstocks. However, specific "energy crops" such as switchgrass and miscanthus can be grown for use in the production of biofuels.

Other agricultural commodities, including jatropha, cassava, and woody crops (short rotation crops: willow, poplar, eucalyptus) are also used as feedstocks for the production of biofuels. Nonagricultural commodities have become a major source for biodiesel production too and used cooking oil is one of them. However, used cooking oil is mainly derived from vegetable oil, which shows the importance of tracing the demand for biofuels to the agricultural market.

This chapter discuss the linkage between the energy-biofuels-agricultural markets, which is very crucial to understand how these markets interact with each other. We explain how the biofuels sector is modeled within the OECD-FAO's[1] Aglink-Cosimo model, which plays a key role in bridging the agricultural and energy sectors.

2. Biofuels market

To understand the linkage between the biofuels and energy markets, it is necessary to understand how the biofuels market works. This market follows a similar structure to any other market, which is consistent with the theory of demand and supply.

On the supply side, biofuels production costs depend on the feedstocks producer price index derived from agricultural markets. On the demand side, biofuels demand is tied to petroleum products demand and the price ratio between biofuels (ethanol and biodiesel) price and gasoline and diesel price. The demand and supply of biofuels interact and solve for the equilibrium of price and quantity (Fig. 4.1). On the supply side, the biofuels market is linked with the agricultural market via feedstocks price, while on the demand side it is linked to the energy market through gasoline and ethanol demands.

---

[1] OECD-FAO: Organization for Economic Cooperation and Development; FAO: Food and Agricultural Organization.

The critical component of the demand side of the biofuels market is the linkage to the energy market. Total gasoline and diesel demand drives the volume of low-blend biofuels use, therefore blended biofuels demand is driven by both the fossil fuel price as well as biofuels price and income. Fig. 4.2 shows how total biofuels use can be subdivided into low-blend and high-blend uses. Low-blend use can either be driven by the market until it hits

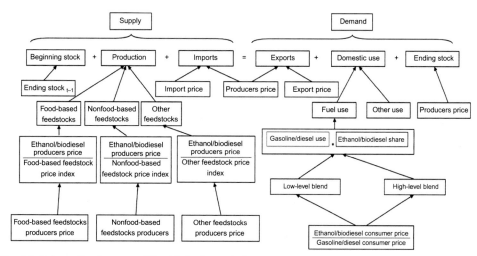

Fig. 4.1 Schematic diagram of the biofuels market.

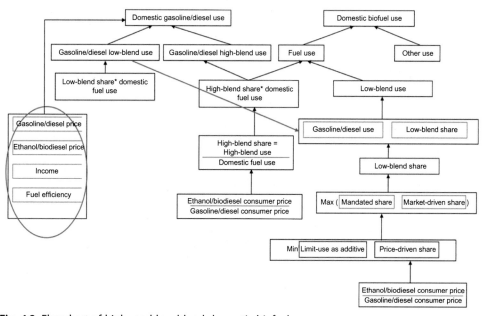

Fig. 4.2 Flowchart of high- and low-blend domestic biofuels use.

the blend wall or be used as a fuel additive. Low-blend biofuels domestic use is driven by policy as well as a major source of fuel additives, while high-blend use is sensitive to biofuel and petroleum product price ratios. Both market-driven low- and high-blend biofuels uses are driven by the relative price of biofuels to the corresponding fossil fuel price.

3. Linkage between energy and agricultural sector

Historically, biofuels price follows the fossil fuel price and feedstocks price. Figs. 4.3 and 4.4 show the price linkage between biofuels (ethanol and biodiesel), petroleum products (gasoline and diesel), and feedstocks (maize and vegetable oil). This is very important to understand the consequence of any movement in one of the markets compared to the other two markets. The lag effect between the changes in the feedstocks price and the changes in biofuels has also been noticed. It justifies the fact that the production in period t depends on the price of the input of the $t-1$ period. The concept goes beyond the biofuels market and can also be applied to any other markets. It is consistent with the theory of "adaptive expectation."

The linkage between energy markets, biofuels, and the agricultural market is depicted in Fig. 4.5. In the Aglink-Cosimo model the links between nonrenewable energy (fossil fuel) and biofuels are unidirectional, that is, the biofuels market considers the changes in gasoline and diesel price and demand, while the energy market does not take into consideration biofuels (ethanol and biodiesel) prices as the proportion of production and consumption of

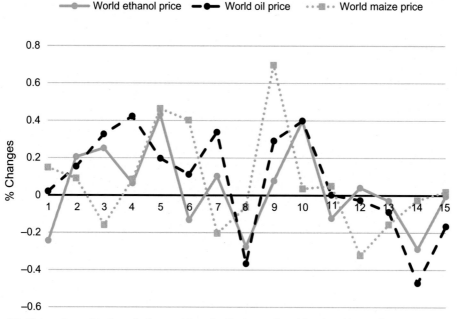

Fig. 4.3 Change in world ethanol price, world crude oil price, and world maize price, % change year-to-year.

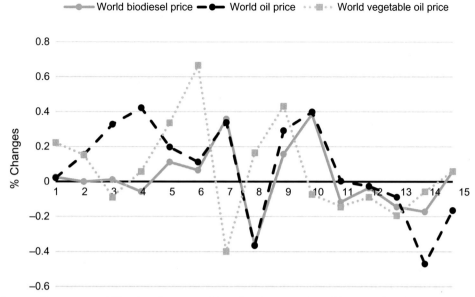

Fig. 4.4 Change in world biodiesel price, world crude oil price, and world vegetable oil price, % change year-to-year.

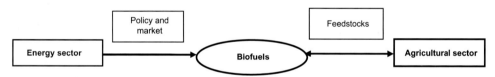

Fig. 4.5 Linkage between energy and agricultural sectors via biofuels.

biofuels are significantly low compared to the production and consumption of fossil fuel. However, with the increase in the use of biofuels, it could be important to consider the two-way linkage between the biofuels and energy markets. The price of biofuels might impact the demand for gasoline and diesel use as a cross-effect. On the other hand, the link between the biofuels and agricultural markets is two way. The changes in agricultural feedstock prices can also lead to changes in the production of biofuels and vice versa. However, the changes in biofuels price have limited or no impact on the changes in gasoline and diesel prices, which is driven by the demand and supply mechanism of the energy sector.

4. Nonrenewable fuel use

Does gasoline and diesel demand depend on ethanol and biodiesel prices? To answer this question, simple regression techniques is used to establish the link between the energy sector and the biofuels sector. Based on the conventional theory of demand, gasoline and diesel use can be estimated as a function of own price (fossil fuel price), cross price (biofuel price), and income effect. In the transportation sector, changes in technology are not uncommon, Vehicles in the 21st century are more fuel efficient than their predecessors. Therefore, in order to capture these technological changes it is important to add a trend term to the gasoline and diesel demand estimation procedure. Among many other estimations, one of the common functional forms is the log–log model as it's coefficients is easily interpreted as elasticities and below is a representation of the country-specific domestic gasoline and diesel demands equation:

$$\ln F_{q,c,t} = k_0 + k_1{}^* \ln CP^F_{q,c,t} + k_2{}^* \ln CP^{BF}_{q,c,t} + k_3{}^* \ln I_{c,t} + k_4{}^* V_{c,t} + \varepsilon^{BF}_{q,c,t} \tag{4.1}$$

where

$F_{q,c,t}$ = the per capita q (gasoline or diesel) type of fuel use in country c and year t
$CP^F_{q,c,t}$ = the consumer price of q (gasoline or diesel) type of fuel in country c and year t
$CP^{BF}_{q,c,t}$ = the biofuels (ethanol or biodiesel) consumer price associated with q (gasoline or diesel) type of fuel use in country c and year t
$I_{c,t}$ = the per capita gross domestic product (GDP) in country c and year t
$V_{c,t}$ = vehicle fuel efficiency in country c and year t (trend)
k_0, and k_1, k_2, k_3, k_4 = the intercepts and slope coefficients
$\varepsilon_{q,c,t}$ = the corresponding error terms.

The endogenous fuel equations play an important role in establishing the linkage between the biofuels and agricultural markets as these equations are used to trace the volatility of the energy market to the agricultural market. Domestic policy dictates complementary biofuels use, which depends on total fuel use. The aforementioned functional form can play an important role in estimating the total complementary ethanol and biodiesel demand.

The use of biofuels is driven by mandates which in most countries are introduced in terms of percentage of total fuel use, it means that any changes in fossil fuel use could change biofuel demand. It is thus important to trace total gasoline and diesel use to the biofuels market. The adoption of these endogenous fuel-use equations in linking the agricultural market to the energy market via the biofuels market can capture both the complement and substitution effects between fossil fuels and biofuels, even though in the broader aspects of the energy market that determines the global crude oil price remains fairly undisturbed.

To capture the substitution effects, the use of a high biofuels ($HB^{BF}_{q,c,t}$) blend in type q fuel in country c and year t can be introduced as an exponential function of biofuels ($CP^{BF}_{q,c,t}$) and fossil fuel ($CP^F_{q,c,t}$) price ratios as shown here:

$$HB^{BF}_{q,c,t} = HS^{BF^*}_{q,c,t} F_{q,c,t}/EE^{BF} \tag{4.2}$$

$$HS^{BF}_{q,c,t} = \frac{1}{\left(1 + \exp\left(4{}^*\kappa{}^*\left(CP^{BF}_{q,c,t}/CP^F_{q,c,t} - EE^{BF}\right)\right)\right)} \tag{4.3}$$

where

$HS^{BF}_{q,\,c,\,t}$ = the high biofuels blend (ethanol or biodiesel) share in type q fossil fuel in country c in year t

EE^{BF} = the energy equivalent of biofuels to fossil fuels

and all other variables defined at an earlier stage.

The substitution of gasoline and diesel by ethanol and biodiesel plays an important role in the context of the linkage between the biofuels sector and the agricultural sector. In the case when global crude oil price rise it give consumers an option to switch to an alternative fuel source upon availability. However, only flex-fuel vehicle owners can take advantage of this option because a conventional car cannot drive on high blends of ethanol. These are the vehicles that can run either on gasoline or on (hydrous) ethanol. While most of the biodiesel can be easily substituted for diesel in vehicles with diesel engines, it is not the case in colder regions where biodiesel derived from vegetable oil freezes at a lower temperature. In developing countries, diesel engine buses are an important mode of public transportation and many people rely on them. Governments are looking for ways to reduce associated GHG emissions by replacing fossil based fuel with biofuels.

5. Policy

Mandates on the domestic use of ethanol or biodiesel play an important role in linking energy and agricultural markets. Indeed, given that policies regulate biofuels demand, the link between biofuels and crude oil prices is relatively limited, except that crude oil prices drive total fuel use, which further drives biofuels use. However, the development of biofuels beyond mandate levels depends on the comparative price ratio between biofuels and crude oil. When the price of crude oil falls, biofuels are less competitive resulting in lower market-driven demand and lower investments in the sector. Higher fossil fuel demand might increase the mandate driven biofuels use which can partially offset the reduction in high blend biofuels use.

Domestic mandates for biofuels can be either binding or nonbinding (Fig. 4.6A and B) depending on the country-specific use of biofuels. The mandate is nonbinding if the mandated level of biofuels use is below the market equilibrium (Fig. 4.6A) and binding when the domestic mandate pushes biofuels use and production beyond the conventional market equilibrium (Fig. 4.6B). In the case of the binding mandate, the biofuels price is above the equilibrium price, while the nonbinding mandate has no effect on market equilibrium. The binding mandate is considered as a policy support to biofuels producers.

6. Interaction between biofuels and agricultural markets

The biofuels market is linked with agricultural markets mainly via the food-based feedstocks demand which includes maize, sugar, wheat, and rice (Table 4.1). Beyond the

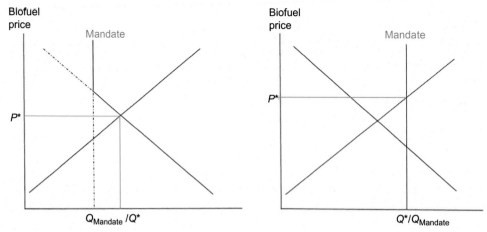

Fig. 4.6 (A) Nonbinding biofuels use. (B) Binding biofuels use.

traditional food-based feedstocks the demand for nonfood-based feedstocks is also important. The inclusion of endogenous nonfood-based feedstock use allows the linkage to
trace direct and indirect land-use changes which also affect the agricultural commodity
market through cross-price effects.

Policies regulate biofuels demand which in turn drives domestic production and
increase feedstock demand. Fig. 4.7 shows the links between the biofuels market and
the agricultural commodity market. The biofuels market is linked through the feedstocks
used in the agricultural commodity markets. The food-based feedstocks used for biofuels
production directly compete with the livestock and dairy industry because an increase
(decrease) in their demand moves the feedstock market and further drives livestock price.
Biofuel serves as a direct link between the energy and agricultural sectors.

Beyond conventional food-based feedstocks several other feedstocks including dedicated energy crops and agricultural residues that result in higher GHG emission reductions
could be important to the overall biofuels market (Table 4.1). Ethanol is generated from the
major food-based feedstocks—maize, sugar cane, and sugar beet and biodiesel is derived
from different types of vegetable oil and used cooking oil and animal fats. Energy
crops—switchgrass and miscanthus, agricultural and forestry residues—can be used to produce either ethanol or biodiesel depending on the biofuel production facilities. The choice
of feedstocks depends on the geographical locations and profit maximizing biorefineries
rely on readily available feedstocks at the lowest price. The consequence of such behavior
is the dominance of sugar cane-based biofuel plants in Brazil, which is the largest sugar cane
producer in the world, while most of the biorefineries extracting ethanol from maize are
located in the Mid-west of the United States. It is not surprising that the United States is the
major maize producer in the world. There are several other feedstocks that can be used in
the production of ethanol and biodiesel, which are listed in Table 4.1.

Table 4. 1 Feedstock use in the production of two types of biofuels: ethanol and biodiesel

Feedstock use	Biofuel	Products included
Maize	Ethanol	—
Molasses	Ethanol	—
Other coarse grain	Ethanol	Barley and rye
Rice	Ethanol	—
Root and tubers	Ethanol	Cassava and potato
Wheat	Ethanol	—
Sugar beet	Ethanol	—
Sugar cane	Ethanol	—
Jatropha	Biodiesel	—
Vegetable oil	Biodiesel	Palm oil, rapeseed oil, soybean oil, sunflower oil, and other vegetable oil
Waste oils and fats	Biodiesel	Used cooking oil and animal fats
Agricultural residue	Ethanol/biodiesel	Wheat straw, bagasse, and corn stover
Dedicated energy crops	Ethanol/biodiesel	Woody crops (short rotation crops: willow, poplar, eucalyptus), and herbaceous crops (perennials: switchgrass, miscanthus)
Forestry residue	Ethanol/biodiesel	Wood chips and logging residues
Municipal solid waste	Ethanol/biodiesel	—
Other waste and residues	Ethanol/biodiesel	Notably paper and pulp industry waste

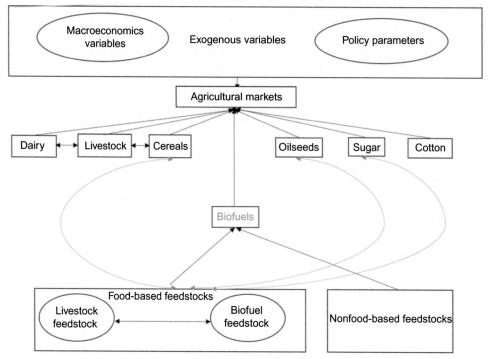

Fig. 4. 7 Linkages between international biodiesel markets and the agricultural, livestock, dairy, and sugar markets.

Mathematical representations are used to establish the linkage between the biofuels and agricultural markets. As there are different feedstocks used for the production of either ethanol or biodiesel, the following equations are used separately to establish the link between ethanol and biodiesel and their corresponding feedstocks market:

$$QP_{c,t}^{ET} = \sum_{i=1}^{n} FS_{i,c,t}^{ET} \tag{4.4}$$

where

QP_{c,t^E} = the total quantity of ethanol produced in country c and year t

$FS_{i,c,t}^{ET}$ = the ith type of feedstock used in the production of ethanol in country c and year t.

$$QP_{c,t}^{BD} = \sum_{i=1}^{m} FS_{i,c,t}^{BD} \tag{4.5}$$

where

$QP_{c,t}^{BD}$ = the total quantity of biodiesel produced in country c and year t

$FS_{i,c,t}^{BD}$ = the ith type of feedstock used in the production of biodiesel in country c and year t.

The main reason behind showing separate equations corresponding to each of the markets is that not all feedstocks can be used to produce biofuels. In other words, biofuels production is tied up with specific feedstocks. Therefore each feedstock used in the production of ethanol and biodiesel is traced separately, using the prescribed equations (4.6) and (4.9):

$$\log FS_{i,c,t}^{ET} = \nu_0 + \nu_1{}^* \log RM_{i,c,t}^{ET} + \nu_2{}^* \log RM_{i,c,t-1}^{ET} + \nu_3{}^* \log RM_{i,c,t-2}^{ET} + \nu_4{}^* \log RM_{i,c,t-3}^{ET}$$
$$+ \nu_5{}^* \log FS_{i,c,t-1}^{ET} + \varepsilon_{FS}^{ET}$$

where   (4.6)

$RM_{i,c,t}^{ET}$ = the profit derived from utilizing ith feedstocks for the production of ethanol in country c and year t.

The $RM_{i,c,t}^{ET}$ is derived based on the following identity:

$$RM_{i,c,t}^{ET} = PP_{c,t}^{ET} + DP_{i,c,t}^{ET} + VL_{i,c,t}^{ET} PI_{i,c,t}^{ET}$$ (4.7)

where

$PP_{c,t}^{ET}$ = the ethanol producer price in country c and year t

$DP_{i,c,t}^{ET}$ = the direct government support for production of ethanol tied with the use of feedstock i in country c and year t

$VL_{i,c,t}^{ET}$ = the value of by-products derived from the use of feedstock i in ethanol production in country c and year t

$PI_{i,c,t}^{ET}$ = the production cost index associated with the use of feedstock i in the production of ethanol in country c and year t

ε_{FS}^{ET} = the corresponding error term.

The following equation is used to derive the production cost index of ethanol for each feedstock use:

$$PI_{i,c,t}^{ET} = SH_{i,c,t}^{ET*} PP_{i,c,t-1}^{ET} + EN_{c,t}^{ET*} OP_{c,t}{}^* ER_{c,t} + CO_{i,c,t}^{ET*} GD_{c,t}$$ (4.8)

where

$SH_{i,c,t}^{ET}$ = the share of feedstock i in the cost index in the production of ethanol in country c and year t

$PP_{i,c,t}^{ET}$ = the producer price index of feedstock i in the production of ethanol in country c and year t in real term

$EN_{c,t}^{ET}$ = the energy share in the cost index in the production of ethanol in country c and year t

$OP_{c,t}$ = the crude oil price index in country c and year t in real term

$ER_{c,t}$ = the exchange rate in country c and year t in real term

$CO_{i,c,t}^{ET}$ = the share of coproducts corresponding to the use of feedstock i in the cost index in the production of ethanol in country c and year t

$GD_{c,t}$ = the GDP deflator in real terms in country c and year t.

The foregoing equations serve as the link between ethanol and agricultural markets for each country-specific feedstocks (agricultural) market.

$$\log FS^{BD}_{i,c,t} = v_6 + v_7{}^* \log RM^{BD}_{i,c,t} + v_8{}^* \log RM^{BD}_{i,c,t-1} + v_9{}^* \log RM^{BD}_{i,c,t-2} + v_{10}{}^* \log RM^{BD}_{i,c,t-3}$$
$$+ v_{11}{}^* \log FS^{ET}_{i,c,t-1} + \varepsilon^{BF}_{FS}$$

(4.9)

where

$RM^{BD}_{i,\ c,\ t}$ = the profit derived from utilizing ith feedstocks for the production of biodiesel in country c and year t

$$RM^{BD}_{i,c,t} = PP^{BD}_{c,t} + DP^{BD}_{i,c,t} + VL^{BD}_{i,c,t} PI^{BD}_{i,c,t}$$

(4.10)

where

$PP^{BD}_{c,\ t}$ = the biodiesel producer price in country c and year t

$DP^{BD}_{i,\ c,\ t}$ = the direct government support for production of biodiesel tied with the use of feedstock i in country c and year t

$VL^{BD}_{i,\ c,\ t}$ = the value of by-products derived from the use of feedstock i in biodiesel production in country c and year t

$PI^{BD}_{i,\ c,\ t}$ = the production cost index associated with the use of feedstock i in the production of biodiesel in country c and year t

ε^{BF}_{FS} = the corresponding error term.

The following equation is used to derive the production cost index of biodiesel for each feedstock use:

$$PI^{BD}_{i,c,t} = SH^{BD^*}_{i,c,t} PP^{BD}_{i,c,t-1} + EN^{BD^*}_{c,t} OP_{c,t}{}^* ER_{c,t} + CO^{BD^*}_{i,c,t} GD_{c,t}$$

(4.11)

where

$SH^{BD}_{i,\ c,\ t}$ = the share of feedstock i in the cost index in the production of biodiesel in country c and year t

$PP^{BD}_{i,\ c,\ t}$ = the producer's price index of feedstock i in the production of biodiesel in country c and year t in real term

$EN^{BD}_{c,\ t}$ = the energy share in the cost index in the production of biodiesel in country c and year t

$ER_{c,t}$ = the exchange rate in country c and year t in real term

$CO^{BD}_{i,\ c,\ t}$ = the share of coproducts corresponding to the use of feedstock i in the cost index in the production of biodiesel in country c and year t

The foregoing equations serve as the link between the biodiesel and agricultural markets for each country-specific feedstocks (agricultural) market specified in Table 4.1.

7. Modeling the market linkage: An example of the OECD-FAO Aglink-Cosimo model

The Aglink–Cosimo model is jointly managed by the Secretariats of the Organization for Economic and Cooperation and Development (OECD) and the Food and Agriculture Organization (FAO) of the United Nations. Each year the Aglink-Cosimo model is used to generate the baseline for biofuels market presented in the *OECD-FAO Agricultural Outlook*. The biofuels model analyses several biofuels–related forward-looking policy scenarios. The ability to capture the interaction between biofuels and agricultural commodities, and across countries is a major strength of this model. It allows policymakers and analysts to track any unintended consequences of biofuels policies on other agricultural sectors.

The biofuels component of the Aglink-Cosimo model is a structural economic model that analyses the world supply and demand for ethanol and biodiesel. The biofuels module is a recursive dynamic, partial equilibrium model used to simulate the annual market balances and price for the production, consumption, and trade quantity of ethanol and biodiesel worldwide. This model is completely integrated with the cereals, oilseeds, and sugar component of the Aglink-Cosimo model. The production of biofuels drives the additional demand for agricultural commodities in particular coarse grains, vegetable oil, and sugar.

The major characteristics of the biofuels module are as follows:
- The ethanol and biodiesel markets worldwide are assumed to be competitive with buyers and sellers behaving as price takers. Domestic and international ethanol and biodiesel prices are determined by solving for the equilibrium demand and supply of each type of biofuel.
- Biofuels (BFL) model is nonspatial, that is, ethanol and biodiesel are assumed to be homogeneous and countries that import ethanol and biodiesel do not distinguish biofuels based on the country of origin.
- Like other components of the Aglink-Cosimo model the BFL module is recursive-dynamic. In each of the projection periods the model is solved based on the prior year outcome. The model simulates 10 years of outcomes into the future which is referred to as the medium-term outlook.
- Ethanol and biodiesel are assumed to be mutually exclusive as such there is no substitution effect between the uses of these two fuels.
- Nonfuel use of ethanol and biodiesel are not modeled and are treated as exogenous.

The BFL model maintains all the linkages mentioned earlier in this chapter and links the biofuels sector to the agricultural sector through the feedstocks market. Even though the Aglink-Cosimo model does not include an energy market model the inclusion of gasoline and diesel demand within the BFL model makes it capable of tracing changes in gasoline and diesel demand that would be driven by macroeconomic factors, including the changes in GDPs and world crude oil prices. This BFL model maintains uniformity across ethanol and biodiesel and across country modules so that biofuels production and use are easily traceable. Within the constraints of this uniformity (templated framework), the BFL model captures individual domestic country-specific mandates, taxes, subsidies, and trade policies.

The BFL model is a part of the Aglink-Cosimo model. Within this framework the demand and supply of a commodity as driven by market conditions, with the exception of a few countries and commodities. However, there is important difference for BFL module. Irrespective to the relative biofuels prices over crude oil prices there are minimum levels of demand for biofuels as they are strongly driven by the policies. The specific use of biofuels is driven by the domestic ethanol and biodiesel use mandates, referred here as domestic mandates. The forward-looking biofuels market outlook developed by the Aglink-Cosimo model relies heavily on country-specific domestic biofuels use mandate assumptions, which further drives the production of biofuels as well as the feedstock demand. In most countries the biofuels use mandate is defined in terms of percentage of the total gasoline and diesel use which makes the inclusion of gasoline and diesel demand important. The BFL model has successfully included the overall country-specific gasoline and diesel demand and estimates the mandate-driven ethanol and biodiesel demand carefully.

The BFL module includes a templated model for the following countries: Argentina, Australia, Brazil, Canada, China, European Union, Japan, Korea, Mexico, New Zealand, Norway, Russian Federation, Switzerland, United States, Colombia, Chile, India, Indonesia, Malaysia, Paraguay, Philippines, and Thailand. Ethanol and biodiesel are modeled separately for each country.

The Aglink-Cosimo model has cereals, oilseeds, and oilseeds products; livestock, dairy, and dairy products; and cotton, sugar, and BFL modules. Each module is linked to the others for example livestock and cereals production competes for land, while herd size in the livestock module depends on the dairy component. The BFL model is linked to the other components of the Aglink-Cosimo model mainly via the food-based feedstocks demand, which clearly is consistent with the linkage theory discussed earlier in the chapter.

8. Equilibrium condition used to solve the biofuels market within the Aglink-Cosimo model

The BFL model has its own market equilibrium for each country covered by the Aglink and Cosimo models. The two market equilibrium conditions that correspond to the specific ethanol and biodiesel markets are represented as follows:

$$ST^{ET}_{c,t-1} + QP^{ET}_{c,t} + IM^{ET}_{c,t} = QC^{ET}_{c,t} + EX^{ET}_{c,t} + ST^{ET}_{c,t} \qquad (4.12)$$

where

$ST^{ET}_{c,\,t-1}=$ the beginning stocks of ethanol in country c and year t

$QP^{ET}_{c,\,t}=$ the production of ethanol in country c and year t

$IM^{ET}_{c,\,t}=$ the import of ethanol in country c and year t

$QC^{ET}_{c,\,t}=$ the domestic use of ethanol in country c and year t

$EX^{ET}_{c,\,t}=$ the export of ethanol in country c and year t

$ST^{ET}_{c,\,t}=$ the ending stock of ethanol in country c and year t.

$$ST^{BD}_{c,t-1} + QP^{BD}_{c,t} + IM^{BD}_{c,t} = QC^{BD}_{c,t} + EX^{BD}_{c,t} + ST^{BD}_{c,t} \qquad (4.13)$$

where

$ST^{BD}_{c,\,t-1}=$ the beginning stocks of biodiesel in country c and year t

$QP^{BD}_{c,\,t}=$ the production of biodiesel in country c and year t

$IM^{BD}_{c,\,t}=$ the import of biodiesel in country c and year t

$QC^{BD}_{c,\,t}=$ the domestic use of biodiesel in country c and year t

$EX^{BD}_{c,\,t}=$ the export of biodiesel in country c and year t

$ST^{BD}_{c,\,t}=$ the ending stock of biodiesel in country c and year t

Based on the foregoing conditions the BFL module determines country-specific ethanol and biodiesel producer prices separately.

The equilibrium conditions that solve the BFL model for world ethanol and biodiesel prices separately are as follows:

$$\sum_{c=1}^{n} EX^{ET}_{c,t} = \sum_{c=1}^{n} IM^{ET}_{c,t} \qquad (4.14)$$

where

$EX^{ET}_{c,\,t}=$ the ethanol export in country c and year t

$IM^{ET}_{c,\,t}=$ the ethanol import in country c and year t

$$\sum_{c=1}^{n} EX^{BD}_{c,t} = \sum_{c=1}^{n} IM^{BD}_{c,t} \qquad (4.15)$$

where

$EX^{BD}_{c,\,t}=$ the biodiesel export in country c and year t

$IM^{BD}_{c,\,t}=$ the biodiesel import in country c and year t

Biofuels are produced from various feedstocks and differs from country to country.

9. Conclusion

This chapter biodiesel the functional form, equations, and variables require in establishing the interaction between the biofuels, energy, and agricultural markets. It discussed the importance of the biofuels market in the context of agricultural commodity markets. One important aspect of this chapter was the discussion of the steps to link the biofuels market to the gasoline and diesel markets which is a key factor in driving policy-driven country-specific domestic biofuels use. On the other hand, market-driven biofuels demand within both low and high blends is also shown. The mathematical representation of the linkage between the biofuels and agricultural markets will help to understand these relationship. The functional forms discussed in this chapter are useful in understanding the practical side of the biofuels market. The proposed gasoline and diesel demand equations with ethanol and biodiesel prices as the cross-effect can be considered as a new way to revisit the relationship between the biofuels and energy markets. The inclusion of high-blend biofuels use and diverse sources of food- and nonfood-based feedstocks make this chapter a valuable tool to develop a conceptual framework for biofuels market interactions.

In the absence of a direct linkage between the petroleum model and the BFL model the inclusion of endogenous nonrenewable fuel equations along with the high-blend ethanol and biodiesel components makes it possible to simulate policies under different sets of macroeconomic variables, including a wide range of crude oil prices, and to trace its consequences to the world agricultural market through biofuels.

Disclaimer

The views expressed are those of the author and do not necessarily represent the official views of the OECD or of the governments of its member countries.

Further reading

OECD/FAO, 2017. OECD-FAO Agricultural Outlook 2017–2026—"Biofuels". chapter, full document available at http://www.agri-outlook.org/.

OECD, 2010. Bioheat, Biopower and Biogas: Developments and Implications for Agriculture.

OECD, 2008a. Biofuel Support Policies—An Economic Assessment.

OECD, 2008b. Developments in Bioenergy Production Across the World: Electricity, Heat and Second Generation Biofuels.

OECD Database of Policies in the Fertilizer and Biofuel Sectors of OECD Countries and Several Emerging Economies. http://www.oecd.org/tad/agricultural-policies/support-policies-fertilisers-biofuels.htm.

Agricultural Market Information System (AMIS) Policy Database Including Biofuel Policies. http://statistics.amis-outlook.org/policy/.

Sector/market integration, contributions, debates and challenges

CHAPTER 5

The food-fuel-fiber debate

Simla Tokgoz
Markets, Trade and Institutions Division, International Food Policy Research Institute, Washington, DC, United States

Contents

1. Introduction

Disclaimer: Any opinions stated in this book are those of the author(s) and are not necessarily representative of or endorsed by IFPRI.

Since the early 2000s, the global biofuels sector has expanded significantly, having important implications for the international agricultural sector. Global ethanol production increased from 33 billion liters in 2002 to 119.7 billion liters in 2016 (OECD/FAO, 2017). During the same period, global biodiesel production increased from 661 million liters to 36.2 billion liters (OECD/FAO, 2017). There were multiple causes behind this fast expansion such as (1) mandates and targets for blending, (2) incentives for biofuels research, and (3) higher gasoline and diesel prices that allowed the blending of biofuels with traditional fuels at a competitive level (National Research Council Report, 2011; Hochman and Zilberman, 2016; Laborde et al., 2016).

However, the 2007–2008 and 2011 food price crises and concern over indirect land use change (ILUC) from biofuels expansion led to further debate on the viability of using agricultural feedstocks to meet transportation fuel demand. Thus, the ongoing biofuels expansion continues to generate discussion in terms of its implications for the global agricultural sector and its ability to meet the world's food, fuel, and fiber needs. This chapter reviews various issues that characterize the food-fuel-fiber debate.

Biofuels, Bioenergy and Food Security
https://doi.org/10.1016/B978-0-12-803954-0.00005-X

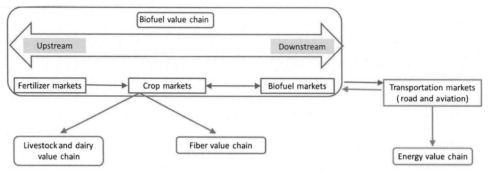

Fig. 5.1 Food-fuel-fiber value chains. *(Source: Author's representation).*

The expansion of the biofuels sector has added new downstream value chains to the existing agricultural feedstocks value chains, and has integrated the agricultural value chain and the energy value chain even more. Fig. 5.1 illustrates the food–fuel–fiber value chains and how they are currently linked. Biofuels expansion has increased the demand for agricultural crops, and created a new downstream value chain for these crops, where they are inputs into production of ethanol and biodiesel or other forms of renewable energy that are being produced using biomass. Higher demand for crops has decreased feedstock availability for the livestock and dairy sector, and has increased input prices for these value chains. At the same time, higher crop prices led to higher crop areas for biofuel feedstocks, such as corn and sugar cane. This leaves less crop area to be allocated to other crops, such as cotton, and having implications for the fiber value chain. Higher crop prices have increased crop production, leading to higher demand for agricultural inputs and affecting upstream value chains, i.e., fertilizer or other inputs. Since biofuels are used as transportation fuels, either as a substitute or as a complement, this expansion has linked energy and agricultural sectors more intensively through transportation markets (Dodder et al., 2015; Elobeid et al., 2013). Prior to biofuels sector expansion, this link was through the use of natural gas in fertilizer production and and the use of diesel and gasoline in farm machinery. To sum up, the biofuels sector has impacted multiple sectors and value chains, leading to an intense debate about competition among food-fuel-fiber needs.

This chapter contributes to the food–fuel–fiber debate by trying to understand the link between these value chains, specifically how biofuels expansion is leading to increased value chain integration, and the relative role of biofuels in each segment of the value chain. The chapter is organized as follows. Next, the current and projected role of biofuels in transportation markets and the energy value chain is discussed. Then, the importance of the biofuels value chain in crop markets is illustrated, specifically through looking at the variety and magnitude of agricultural feedstocks utilized in biofuels production. Subsequently, the impact of biofuels expansion on the livestock value chain and fiber value chain is explored. Consequently, the state of food-fuel-fiber debate is reviewed, with a summary of the main points raised in the literature. Conclusions follow with some ideas on the way forward.

2. Role of biofuels in the energy value chain

As presented in Fig. 5.1, the biofuels sector is part of the energy value chain, through its use in the transportation sector. In the beginning, many countries utilized biofuels as drop-in fuel in addition to traditional fossil fuels, such as E10 blends. Later, the role of biofuels expanded to being an alternative fuel in markets like Brazil for flex-fuel vehicles (FFVs) (Balcombe and Rapsomanikis, 2008) and the United States for E85 vehicles (Anderson, 2012). Use of biofuels in aviation is new but growing (Tyner and Petter, 2014).

Biofuels expansion represents part of the global trend toward renewable forms of energy. Biofuels as transportation fuel, albeit having a significant impact on the agricultural sector, constitute only a small part of the renewable energy sector. To describe the relative status of biofuels in traditional and renewable energy sectors, let us examine the International Energy Agency's World Energy Outlook (WEO) (2016) that includes historical data and projections for the global energy sector.

WEO (2016) presents a diverse set of trends across regions and across energy sources for global energy sector in three scenarios: New Policies, Current Policies, and 450.[1] In the New Policies Scenario, which is the main scenario in WEO (2016), world primary energy demand is projected to increase 31% between 2014 and 2040. The source of this demand growth is slated to change over the coming decades: a shift from developed countries to fast-growing developing countries is being projected. OECD countries' share in world energy demand is projected to decrease from 39% in 2014 to 28% in 2040. Share of non–OECD countries, in the same period, is projected to rise from 59% to 68% because of industrialization and urbanization (WEO, 2016). Asia, as a region, leads this share increase, followed by Africa, the Middle East, and Latin America. Thus, which energy sources will be used to meet the increasing demand of developing countries is at the heart of the debate.

When we examine the energy resources needed to meet this demand in global terms, the share of coal and oil is projected to decline between 2014 and 2040, whereas the share of nuclear and gas is projected to increase in the same time period. The share of renewables (hydro, bioenergy, and other) is projected to increase as well, from 14% in 2014 to 19% in 2040 globally. In 2040, out of 17,866 million tonnes oil equivalent (mtoe) total primary energy demand, 3,456 mtoe is projected to be met by hydro, bioenergy, and

---

[1] New Policies Scenario contains existing energy policies and an assessment of the results likely to stem from the implementation of announced intentions, particularly those in the climate pledges submitted for COP21. The Current Policies Scenario includes only those policies firmly enacted as of mid-2016. The 450 Scenario shows a pathway to limit long-term global warming to 2°C above preindustrial levels.

other renewables. Renewables are gaining ground due to a variety of factors, such as policy support and declining costs. WEO (2016) estimates that subsidies to renewables are around $150 billion currently, with 80% going to the power sector, 18% to transport, and around 1% to heat. The fastest growth is for the use of renewables for power generation in the Outlook. Use of bioenergy and other renewables in power generation is projected to increase from 9% in 2014 to 41% in 2040 globally. This is due to many factors, including the electricity sector being a focus in COP21, the rapidly declining costs for solar and wind power, and the expansion of subsidies (WEO, 2016).

The role of biofuels is projected to grow as well in the coming decades, but not as fast as the other forms of renewable energy. The transport sector's demand for traditional fossil fuels continues, despite biofuels' increasing their share in transport fuels in the projection period. The transport sector also forms the source of a significant share of global energy demand, making up approximately 19% of global total energy demand throughout the projection period of 2014–2040. Transport demand for energy is projected to grow by 31% between 2014 and 2040 globally in the New Policies Scenario. Biofuels forms a small part of this sector globally, with the share of biofuels projected to increase from 3% in 2014 to 6% in 2040. The use of biofuels in transport is projected to increase from 1.6 mboe per day in 2014 to 4.2 mboe per day in 2040. Of this, the majority will go to road transport (4 mboe per day), with aviation and maritime use increasing slightly (0.25 mboe per day in 2040). Thus it can be observed that even though biofuels use in the transport sector is projected to increase, this growth is not as fast as other forms of renewable energy.

Regional projections for energy demand by the transport sector and utilization of biofuels in transport illustrate that the picture is not uniform globally. OECD countries' energy demand for transport is projected to decrease by 18% between 2014 and 2040, with the share of biofuels projected to increase from 4% to 10% (WEO, 2016). Non-OECD countries, on the other hand, are expected to increase their energy demand for transport by 75% in the same projection period, although the share of biofuels is projected to be low in this region's transport sector: 2% in 2014 and 5% in 2040. In particular, Eastern Europe/Eurasia have low growth in energy demand for transport with a low share of biofuels. Non-OECD Asia, such as China and India, shows 108% growth in energy demand for transport, with the share of biofuels projected to increase from 1% in 2014 to 4% in 2040. These countries are the main drivers of growth in non-OECD countries for the transport sector's energy demand. Middle Eastern countries utilize no biofuels in the transport sector and African countries utilize negligible amounts. Latin America, already including an established biofuels sector with Brazil leading the way, also shows significant growth. Energy demand for transport in this region is projected to increase by 36% between 2014 and 2040, with the share of biofuels doubling to 20% in 2040 from 10% in 2014.

Fig. 5.2 presents historical data and projections for biofuels use in transport across regions, demonstrating different growth paths. OECD Americas, leading the way, is

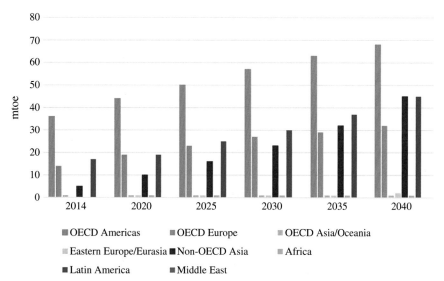

Fig. 5.2 Biofuels use in transport in mtoe. *(From World Energy Outlook, 2016. International Energy Agency, World Energy Outlook 2016, OECD/IEA, Paris, France).*

projected to use 68 mtoe in 2040, followed by non-OECD Asia and Latin America at 45 mtoe each. OECD Europe is projected to consume 32 mtoe biofuels in the transport sector in 2040, whereas the remaining regions are projected to use negligible amounts.

In many countries, government targets and mandates, and other supporting policies, drive growth in the use of biofuels and other renewable energy in the transport sector. WEO (2016) notes that advances in technology, along with subsidies, are leading to this global expansion of biofuels use in transport. REN21 (2017) reports that currently, 43 countries have targets for renewable energy to be used in the transport sector. When specific mandates are considered, 32 countries have blending mandates for ethanol (ranging from E2 to E22.5), and 20 countries have biodiesel blending mandates (ranging from B0.5 to B20) (REN21, 2017). The other forms of support include tax exemptions or cuts, subsidies or grants to develop new technologies (REN21, 2017).

As seen earlier, the transport sector's demand for traditional fossil fuels is still strong. Current global policy framework does not envisage a big change to existing infrastructure. Biofuels are still mainly used as a drop-in fuel in many countries. One alternative to traditional fossil fuel use in transport, electric vehicles (EVs) are increasing on roads. World sales of EVs grew by 70% in 2015 and, for the first time, numbers on the road exceeded 1 million (reaching nearly 1.3 million) (WEO, 2016). REN21 (2017) reports that the EV passenger car market represented approximately 1% of global passenger car sales in 2016. The top five countries for total passenger EV deployment in 2016 were China, the United States, Japan, Norway, and the Netherlands. However, these types of changes do not lead to major overhauls of the transportation infrastructure, but focus on improving an existing system. REN21 (2017) reports that sustainability efforts for the transport sector include more than biofuels in road, aviation, and maritime transport, specifically increasing energy efficiency and use of EVs as two additional means.

To sum up, biofuels use in the global transportation sector is small, but not negligible. At the same time, countries are looking to diversify energy sources for transport, such as EVs. Various forms of renewable energy are being supported, with the aims of diversifying energy sources and achieving environmental goals. Although the biofuels sector is critical for the agricultural sector, its role in renewable energy and traditional energy sectors remains minor.

3. Role of biofuels in the crop value chain

Energy and crop value chains were connected before the expansion of biofuels, through the use of energy in agricultural machinery on farms and in fertilizer production, and transportation costs of agricultural feedstocks (Dodder et al., 2015). The biofuels sector was added to this link by creating a new demand for crops and thus expanding the downstream value chain of agricultural feedstocks.

Food and feed demand constitute the most significant global share of demand for crops. With demand from the biofuels sector expanding, these shares have declined, but still comprise a considerable portion. Let us first explore the demand dynamics for some of the feedstocks used in ethanol production. In 2013, feed demand constituted 60.1% of global corn demand, where it declined from 70.5% in 2000 (FAOSTAT, 2018). Wheat food demand was 74.1% of total demand globally in 2013, a decrease from 78.3% in 2000 (FAOSTAT, 2018). Sorghum feed demand share was 53.8% in 2000 and went down to 45.4% in 2013 globally. For all of these grains, the Other Uses category in the FAOSTAT database increased as a share of total demand in the same time period. For vegetable oils used in biodiesel production, an analysis of the data paints a similar picture. The share of food demand in palm oil demand decreased from 46.3% in 2000 to 30.2% in 2013 (FAOSTAT, 2018). For soybean oil demand, this decline in share is from 79.2% to 58.2%, a very significant drop. Global food demand share of rapeseed and mustard seed oil was 65% in 2000 and 41.9% in 2013. A similar downward trend is observed in sunflower seed oil as well. As seen in the FAOSTAT data, although food and feed use of agricultural feedstocks are still a key demand category, demand from the biofuels industry has significantly expanded, creating a new downstream value chain of substantial size.

This rather fast and sizeable expansion has generated the food-fuel-fiber debate, given the large impact it had on international agricultural markets, and led to an extensive literature analyzing its impacts on consumers and producers in all parts of the value chain. The food-fuel-fiber debate focuses on the impact of this demand growth on the availability of agricultural feedstocks, given the limited potential to increase supply due to limits on global arable land.

There is a wide range of feedstocks that are currently being used for biofuels production—some agricultural and some not. An analysis conducted by the Environmental Protection Agency (EPA) (EPA, 2016) to estimate the "Lifecycle Emissions for Select Pathways" shows the wide range of feedstocks available in the United States for the production of biofuels and the different technological processes that can be employed. The report shows that biodiesel can be produced from nonagricultural sources such as algal oil and yellow grease, and from agricultural feedstocks such as canola oil, palm oil, and soybean oil. Palm oil can also be used for renewable diesel through a different technology, and switchgrass and cellulose from corn stover can be used for producing cellulosic diesel. For ethanol production, possible agricultural feedstocks are barley, corn starch, grain sorghum, and sugar cane. Switchgrass and cellulose from corn stover are other agricultural feedstocks that can be used for ethanol production, but not for food or fiber. Corn starch can be used for butanol production with a different technology. The EU Commission RED Directive of 2009 (EU Official Journal RED-Directive 2009/28/EC, n.d.) conducted a similar analysis and listed rapeseed oil and sunflower oil as agricultural feedstocks, and waste vegetable or animal oil as nonagricultural feedstocks for biodiesel production in the European Union. Sugar beet is also added for potential agricultural feedstocks for ethanol production in the European Union. Table 5.1 presents a list of the currently utilized agricultural feedstocks by major producers of biofuels.

Figs. 5.3–5.5 present the quantity of agricultural feedstocks used in ethanol and biodiesel production for major producers and major feedstocks to illustrate the speed with which the biofuels sector expanded. Fig. 5.3 shows that U.S. grain use for ethanol

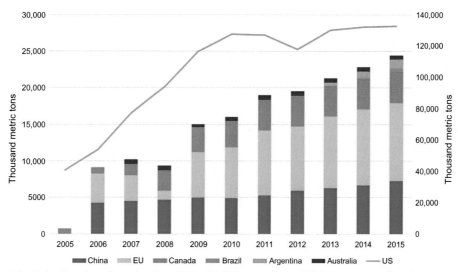

Fig. 5.3 Grains in ethanol production. *(From GAIN Reports, 2018. U.S. Foreign Agricultural Service GAIN Reports, various issues, Available Online at https://gain.fas.usda.gov/Pages/Default.aspx).*

Table 5.1 Agricultural feedstocks used in biofuel production

Commodity	
Ethanol	
Wheat	China, EU, Canada, Australia, Turkey
Corn	US, China, EU, Canada, Turkey, Brazil, Argentina
Barley	EU
Sorghum	Australia, China
Sugar cane	Brazil, Colombia, Thailand, Philippines
Molasses	Argentina, India, Pakistan, Indonesia, Philippines, Thailand
Cassava	China, Thailand
Rye	EU
Biodiesel	
Soybean oil	US, Argentina, EU, Brazil, Canada
Rapeseed oil	EU, Canada, US
Palm oil	EU, Malaysia, Indonesia
Sunflower oil	EU

Notes: The list is not comprehensive and includes major producers. The list focuses on first-generation biofuels.
From GAIN Reports, 2018. U.S. Foreign Agricultural Service GAIN Reports, various issues, Available Online at https://gain.fas.usda.gov/Pages/Default.aspx.

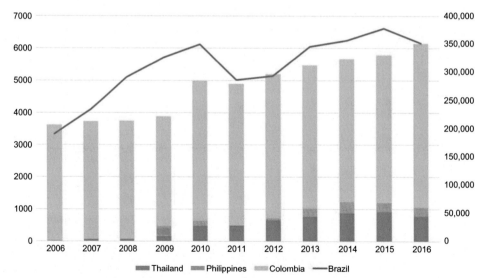

Fig. 5.4 Sugar cane in ethanol production. *(From GAIN Reports, 2018. U.S. Foreign Agricultural Service GAIN Reports, various issues, Available Online at https://gain.fas.usda.gov/Pages/Default.aspx).*

increased by 226% between 2005 and 2015 (an increase of 91,956 thousand metric tonnes [tmt]), becoming a major source of corn demand (GAIN Reports, 2018). For the same time period, this growth rate was 516% in Canada. For China, the growth rate of grains in ethanol production was 71% between 2006 and 2015. For the same period, this growth rate

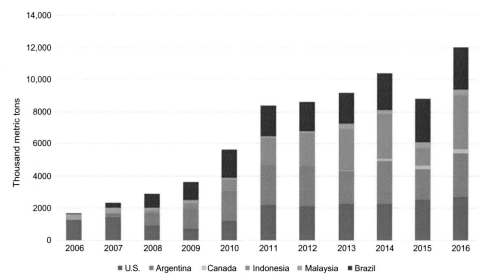

Fig. 5.5 Vegetable oils in biodiesel production. *(From GAIN Reports, 2018. U.S. Foreign Agricultural Service GAIN Reports, various issues, Available Online at https://gain.fas.usda.gov/Pages/Default.aspx).*

was 165% for EU-28 (GAIN Reports, 2018). Fig. 5.4 presents sugar cane use in ethanol production. The Brazilian ethanol sector, already established, increased its use of sugar cane by 85% between 2006 and 2016. In the same period, Colombia increased its use by 42%. For the countries included in Fig. 5.4, total sugar cane use increased by 164,130 tmt between 2006 and 2016 (GAIN Reports, 2018). Vegetable oil use in biodiesel production, presented in Fig. 5.5, demonstrates high growth rates, partially due to very low use in the beginning. We see that vegetable oil use increased by 1,469 tmt (117%) in the United States, 2,683 tmt (15782%) in Argentina, 4,375 tmt (92%) in EU-28, 269 tmt in Canada, 3,337 tmt in Indonesia, and 2,570 tmt in Brazil between 2006 and 2016. In Malaysia, use of vegetable oil declined slightly (GAIN Reports, 2018) during this period.

These figures indicate that, despite efforts to diversify to nonagricultural feedstocks, for many countries the quantity of agricultural feedstocks used in biofuels production is significant and increasing. As seen in the EPA (2016) study and EU Commission RED Directive of 2009, there are potential nonagricultural feedstocks that can be utilized for biofuel production. Not all of these technologies and feedstocks are currently commercially viable. However, they paint a picture of the potential, showing that it is possible to avoid food-fuel-fiber competition through the use of new technology.

4. Livestock and fiber value chains

As illustrated in Fig. 5.1, crop value chains are linked to livestock, dairy, and fiber value chains. Livestock and dairy sectors are important customers in agricultural feedstock markets, generating a large feed demand source globally. Expansion of biofuels value chains has impacted this link, creating competition for agricultural feedstocks with the livestock and dairy sectors.

The main impact of biofuels expansion on livestock and dairy producers has been through higher agricultural feedstock prices, which have decreased the producers' profit margins. This, along with reduced availability of grains, has decreased production and increased meat and dairy prices (Hayes et al., 2009). One factor easing the constraint on livestock and dairy sectors is the use of by-products generated during biofuels production that are inputs to the livestock and dairy industries (e.g., dried distillers grains [DDG] from wheat–, corn–, sorghum–, and sugar beet-ethanol). For example, in the United States, DDG are by-products of ethanol production from corn or sorghum in dry mills (BRD, 2009). Thus, with the expansion of the ethanol sector, DDG production increased 1559% between 2000–2001 and 2016–2017 marketing years, an increase of 34.9 million metric tonnes (mmt) (ERS, 2018). U.S. exports of DDG increased by 10 mmt in the same period, showing that the majority of production in the United States went into domestic utilization. This eased the trade-off between food and fuel purposes in the United States, but it did not eliminate the problem. The availability of DDG is limited compared to other grains providing energy (corn, sorghum, wheat, barley), and cannot completely overcome the effects of biofuels on the livestock and dairy sectors. Most of DDG in the United States are fed to beef cattle, followed closely by dairy cattle (Wisner, 2015). Use in swine and poultry production has been very limited (Wisner, 2015). The National Research Council Report (2011) also noted that use of grains as feed in nonruminants was more limited than ruminants, making their producers more susceptible to price increases with less flexibility to change the diets of their animals. The use of DDG in animal feed to replace grains for energy is limited because farmers can add only a certain amount before running into production and quality issues (BRD, 2009; National Research Council Report, 2011). By-products from the wet mill process to produce ethanol from corn are corn gluten feed, corn gluten meal, and corn oil, which can be used as feed as well (National Research Council Report, 2011; BRD, 2009).

The expansion of biofuels has had an impact on the fiber value chain through the competition for productive agricultural land, specifically between crops used as feedstocks in biofuel plants and crops used for fiber purposes. Higher demand for biofuel feedstocks, such as corn, vegetable oil, and sugar cane, has increased prices of these crops, encouraging farmers to produce more of them. This is accomplished either through increasing yields using inputs more intensively or through increased area allocation. Crop area reallocation to accommodate this higher biofuel feedstock demand has negatively impacted the planted area of crops such as cotton. At the same time, demand for growth of food, feed, and fiber is still robust because of population growth and changing dietary patterns due to income growth (Laborde et al., 2016). Thus overall demand for agricultural land has increased, leading to conversion of land from other uses to agricultural production (ERS, 2011).

5. State of the debate on food-fuel-fiber

5.1 Main determinants and effects of biofuels expansion

The global biofuels sector has expanded considerably. There are various factors contributing to this expansion and at the same time there are many factors limiting the scale of this expansion. Fig. 5.6 summarizes the main determinants and effects, which are discussed in detail next.

In the beginning, the main driver for the biofuels sector expansion was the desire for energy security through diversification of a country's energy sources. Additionally, the desire to find new markets for agricultural crops and increase rural incomes provided impetus for biofuels sector expansion. This expansion was first led by policy frameworks, such as the Energy Policy Act of 2005 in the United States and the EU Biofuels Directive of 2003. Market conditions, such as increases in world crude oil prices starting in 2005 and lasting until 2014, have also made the blending of biofuels in transportation fuel relatively competitive with respect to gasoline and diesel. However, biofuel usage growth was limited by market access constraints in transport fuel markets. There is a blending wall for the blending of ethanol with gasoline and biodiesel with diesel, due to regulatory and infrastructure rigidities that restrict how biofuels can be consumed (Pouliot and Babcock, 2014; Babcock and Pouliot, 2013). For example, both EU-28 and U.S. fleets have E10 as the most common standard. FFVs and E85 vehicles are limited in many countries. For example, in the United States in 2014, total E85 fuel consumption was 19,743 thousand gasoline equivalent gallons (geg), a small share in the market in which

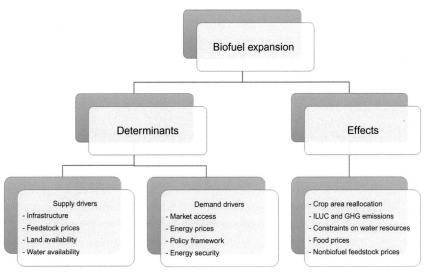

Fig. 5.6 Determinants and effects of biofuels sector expansion. *GHG*, greenhouse gas; *ILUC*, indirect land use change. *(Source: Author's representation).*

motor gasoline consumption, including E10 and E85, is at 136,756,200 thousand geg (AFVD, 2017; EIA, 2017). For biodiesel, most blends do not exceed 8% in EU-28. U.S. biodiesel consumption in 2014 was 1,417 million gallons, with total distillate fuel oil consumption including biodiesel at 61,892 million gallons (biodiesel made up approximately 2% of transportation fuel).

In terms of the supply drivers of biofuels, feedstock prices and availability are critical for the profit margins of biofuel plants. At the same time, the transportation costs of feedstocks to biofuel plants affect the profit margins and the location of biofuel plants (as seen in the majority of U.S. plants being located in the Mid-west). The availability of land and water resources (as well as other inputs to agricultural production) also impact the prices of feedstocks that plants purchase, and thus their profit margin.

In terms of effects from this expansion, first, higher demand for biofuel feedstocks has increased the prices of these crops, causing competition among crops for productive agricultural land. This has resulted in crop area reallocation, leading to higher crop areas for feedstocks used in biofuels production and lower areas allocated to other crops. Second, in some countries, land from other uses has been converted to agricultural production, leading to the ILUC and GHG emissions debate for biofuels (Searchinger et al., 2008; Laborde and Valin, 2012). Higher agricultural production also put constraints on other inputs, such as water resources, fertilizer, and seeds. Furthermore, higher crop prices have led to higher feed prices, decreasing profitability in the livestock and dairy sectors. Thus, food prices have increased due to higher agricultural crop, livestock, and dairy prices. With increasing population and income growth, demand for food and feed remains strong globally, with low food and feed demand elasticity to prices. These effects and their respective role in the food-fuel-fiber debate are discussed next.

5.2 Reaction to biofuels due to food prices

The food price crises of 2007–2008 and 2011 (Fig. 5.7) have triggered intense debate regarding contributing factors. Recent biofuels expansion has received considerable interest in this context, and many empirical studies have been conducted to understand the impact of the biofuels sector on the agricultural sector, and its role in food price increases and agricultural production (Zhang et al., 2013; Serra and Zilberman, 2013; Condon, Klemick, and Wolverton, 2015).

The impact of this expansion on the agricultural sector was through two channels. One was the increase in demand for agricultural feedstocks through the biofuels sector. The other was the increased impact of the energy sector and energy prices on the agricultural sector, since a larger portion of the demand for agricultural feedstocks was now linked to developments in the energy sector.

A sizeable literature has attempted to explore the links between agricultural, biofuels, and energy sectors and has utilized time series econometric techniques to analyze the

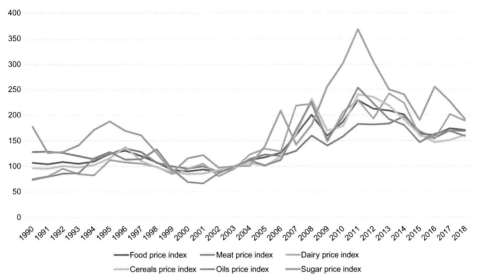

Fig. 5.7 Food and Agricultural Organization food price indices. *(From FAOSTAT, 2018. Database. Available Online at http://www.fao.org/faostat/en/#home).*

price levels of agricultural, biofuels, and in some cases energy commodities, as well as the volatility (and transfer of volatility) of these prices. Serra and Zilberman (2013) provided an extensive review of the literature and concluded that, in terms of price levels, energy prices drive long-run agricultural prices. In terms of volatility transfer, the literature finds evidence supporting this, especially after the mid–2000s (Serra and Zilberman, 2013). When we look at the recent literature, de Nicola et al. (2016) showed that there is high correlation between the price returns of energy and agricultural commodities. The study also showed that the overall level of comovement among energy and agricultural commodities has increased in recent years, specifically in the cases of corn and soybean oil. Lucotte (2016) also found strong positive comovements between crude oil and food prices after 2007, while no statistically significant comovements are observed prior to 2007. Avalos (2014) also found evidence that the transmission of oil price innovations to corn prices became stronger after 2006 (but did not find evidence with respect to soybeans). Therefore it can be concluded that the energy sector's influence on the agricultural sector has been growing, tying agricultural commodity and food prices to energy prices through the demand side in addition to the input side.

Another sizeable literature has focused on the utilization of simulation models and scenario analyses to understand the effects of biofuels and energy sectors on agricultural markets. This literature used partial-equilibrium (PE), linear programming (LP), or computable general equilibrium (CGE) models, which generate projections into the future for supply, demand, and prices. Studies have analyzed the impact of biofuels by generating scenario projections and comparing them to a baseline. These scenarios follow the ceteris paribus principle, and introduce a policy or another type of shock in scenario to understand its implications.

The estimates of the impact of biofuels on agricultural feedstock prices from these simulation models have a wide range in value, and efforts have been made to understand the root of these wide range of estimates, especially to guide the policy debate on biofuels. Zhang et al. (2013) reviewed this literature, specifically studies using PE or CGE models to understand the reasons behind the range in estimated price impacts. The authors identified several potentially important differences among these studies and models. To do this comparison, they focused on biofuels from corn, sugar cane, and vegetable oils, and empirical studies using global models only. For studies using PE models, the range in estimates stemmed from differences in the design of scenarios, crude oil price projections used in the model (these are exogenous in PE models), the presence or absence of biofuel trade in the model, and the structural way in which agricultural and energy market linkages are modeled. For studies using CGE models, the differences were because of model assumptions on agricultural land supply, the inclusion (or lack) of by-products, assumptions on crude oil prices, and the elasticity of substitution between petroleum and biofuels. Zhang et al.'s (2013) comparison among selected PE studies showed that the projected impact of biofuel growth in world prices ranged from 14.6% to 52.6% for corn, from 3.4% to 37.1% for sugar cane, and from 0.4% to 15% for vegetable oils. They found that CGE model results showed a wider range for price increases due to biofuel growth: U.S. corn price increases between 4.7% and 45.2%, Brazil sugar cane price increases between 1.5% and 83.7%, and EU oilseeds price increases between 5.5% and 62.5%.

Hochman and Zilberman (2016) also carried out a systematic review of the literature and conducted a meta-analysis of empirical results, but focused on U.S. corn ethanol and its impacts on GHG emissions, food and fuel prices, and other economic welfare indicators. They reported that the average increase in U.S. corn price in the literature was 61 cents per bushel, a moderate value, with a standard deviation of 65 cents per bushel. However, they acknowledged that their analysis showed that biofuels' impact is higher in the short-run and was an important contributor to the 2007–2008 food price crisis. Similar to Zhang et al. (2013), they drew attention to the heterogeneity of results in the literature, identifying the sources of heterogeneity as (1) inclusion of petroleum markets, (2) whether demand and supply elasticities are inelastic or not, and (3) calibration year of the model.

Although the impact of the biofuels sector on agricultural feedstock prices has been established, it should be noted that there are other factors contributing to food price increases in both the short and the long- term. One study that draws attention to this is the National Research Council Report (2011), which conducted an extensive review of literature on U.S. biofuels policy and its implications. One of the key findings of the report is on food and feed prices, where they wrote "Food-based biofuel is one of many factors that contributed to upward price pressure on agricultural commodities, food, and livestock feed since 2007; other factors affecting those prices included growing population overseas, crop failures in other countries, high oil prices, decline in the value of the U.S. dollar, and speculative activity in the marketplace." Other factors contributing to food price increases have also been described in the literature such as income growth in developing countries, trade restrictions through policy, and historically low grain stocks (summarized by Condon et al., 2015).

The National Research Council Report (2011) made an important distinction between food price increases and agricultural crop price increases. Since the share of crop prices in the costs of food production is low and since marketing margins (for crops such as corn, wheat, and soybeans) are high, food price increases have been at a lower scale than agricultural crop price increases (National Research Council Report, 2011). The impact of agricultural crop price increases on on livestock products is higher since these crop prices constitute a higher share of the cost of production for livestock products (National Research Council Report, 2011).

5.3 Natural resource constraints

The critical driver of the food-fuel-fiber debate is the scarcity of natural resources and how best to utilize them. The prevalent use of first-generation biofuels that directly rely on agricultural feedstocks has intensified this debate and shaped the discussion regarding the future of policy support for biofuels. The production of agricultural feedstocks requires land, water, agricultural inputs (fertilizer, seeds, etc.), and energy inputs (diesel in farm machinery, natural gas in fertilizer). The fast growth of ethanol and biodiesel, combined with other long-term sources of demand growth, has put an immense pressure on the available resources.

Faced with rapid demand expansion and the ensuing price increases, farmers initially increased crop areas allocated to biofuel feedstocks to meet this demand and decreased land allocated to other crops (ERS, 2011). For example, between 2002 and 2012, Brazilian sugar cane areas increased by 90%, as opposed to a 21% increase between 1992 and 2002 (FAOSTAT, 2018). U.S. corn area expanded by 26% between 2002 and 2012, whereas it decreased by 4% from 1992 to 2002 (FAOSTAT, 2018). Global demand for food and feed has remained robust despite higher crop prices due to low food and feed demand elasticities. Conversion of land from other uses into agricultural production has increased (ERS, 2011), having consequences on GHG emissions, water use and quality, and biodiversity. For example, the area under permanent crops increased by 17% between 2002 and 2015 globally (FAOSTAT, 2018).

Even with crop area expansion, it is not possible to meet the biofuel demand foreseen by mandates and targets in many countries. One illuminating example comes from Poudel et al. (2012) who emphasized land constraint in meeting the growing worldwide biofuel demand. The authors computed possible global ethanol production from all available grains and sugar feedstocks for the year 2007 and found that (in oil equivalent terms) this would meet only 19.4% of total transportation gasoline demand. The same computation was done for vegetable oils and biodiesel, and they found that biodiesel would be equal to 7.5% of world oil demand in 2007. Thus the potential for biofuels to meet the growing demand for transportation energy remains limited due to land supply constraints.

When discussing land constraints, it is important to touch briefly on the ILUC debate. One of the main factors of biofuel policy support in the beginning was the fact that biofuels, when used as transportation fuel, led to lower GHG emissions. However, this argument has been countered in recent years with the discussion on ILUC leading to GHG emissions through the release of carbon sequestered on land converted to the production of biofuel feedstocks or through the release of carbon stored in soils and vegetation. These changes occur when, through the impact of higher crop prices caused by higher biofuel demand, demand for agricultural crop areas increases either for traditional crops or for dedicated energy crops. Crop area replaces land uses such as pasture land, forest land, or idle/degraded land.

There are a lot of studies dedicated to simulating the magnitude of ILUC and related GHG emissions (starting with Searchinger et al., 2008; Fargione et al., 2008). A summary of this literature can be found in Tokgoz and Laborde (2014). These studies use PE models such as CARD-FAPRI of ISU (Dumortier et al., 2011), CGE models such as GTAP (Keeney and Hertel, 2009; Tyner et al., 2010) or MIRAGE (Laborde and Valin, 2012), or LP models such as FASOM by Beach et al. (2012) and BEPAM by Chen et al. (2012). Many of these studies have focused on implications of legislative processes on ILUC and GHG emissions (California Low Carbon Fuel Standard, EU Commission RED Directive, UK Renewable Transport Fuel Obligation, U.S. Energy Independence and Security Act of 2007), with some focusing on energy prices' impacts on biofuels and ILUC. An important study in this debate is by Khanna and Zilberman (2012) which summarized the issues critical for modeling land use and GHG emissions through simulation models.

The food-fuel-fiber debate has mostly centered on land availability in the short-term, but land intensification is just as critical to increase supply of agricultural feedstocks. Although global arable land is fixed, yield growth remains a viable solution to increase crop production. Some key solutions to aid yield growth include higher prices allowing farmers to use more inputs, investments in mechanization and irrigation, better land management to increase quality of land, agricultural R&D, use of new crop varieties, and utilization of cropping intensity (Laborde et al., 2016; Poudel et al., 2012; ERS, 2011).

Yield growth on existing crop land and yield levels on newly converted land are both critical. While discussing the potential of yield growth to ease supply constraints, it should be noted that there has been some debate on whether a yield plateau has been reached for crops, especially in developed countries. The maximum level that a crop yield might reach depends on a plant's capacity to capture sunlight and its ability to convert it into biomass and partition it into grain (ERS, 2011). Studies that have concluded that a yield plateau has been reached include Brown (1994), Oram and Hojjati (1995), Pingali et al. (1997), Calderini and Slafer (1998), and Cassman et al. (2003). Alternatively, Reilly and Fuglie (1998) disputed the existence of a yield plateau in the United States. Egli (2008) also argued that there is no convincing evidence that U.S. yields are reaching plateaus in the case of soybeans. Menz and Pardey (1983) also claimed that no yield plateau has yet been reached for corn. It is important to make a distinction between developed and developing countries in this regard: developing countries are further away from their yield plateaus and have the potential to increase supply through higher use of inputs, mechanization, and expansion of irrigated areas in the short term.

5.4 Potential of advanced biofuels

There are research efforts on and experimental use of second-generation biofuels, which are seen as a solution that could ease the constraints of the food-fuel-fiber competition. Materials considered for advanced biofuels development include agricultural residues, municipal wastes, forestry residues, and energy crops (switchgrass and miscanthus). Different technology pathways are used that produce a range of biofuels such as ethanol, butanol, methanol, and diesel jet fuel (IRENA, 2016; REN21, 2017).

Advanced biofuels will also generate consequences such as land use change if crops like miscanthus or switchgrass are used, or negative impacts on soil fertility if agricultural residues like corn stover are used. Use of dedicated energy crops for advanced biofuels will continue to compete with food and feed supply chains for agricultural inputs, most notably land. Higher per-acre ethanol yields from biomass could keep land demand down relative to a similar scale of corn ethanol production. However, these two pathways are not optimal solutions to the food-fuel-fiber competition.

At the same time, utilization of wastes and forestry residues eliminates the competition for crop area and inputs. Furthermore, GHG emissions through ILUC effects are avoided. Avoidance of crop area competition disconnects biofuels expansion from the food supply and food prices debate. This illuminates the need for balance among different technology pathways in policy design and policy support.

There are continuing efforts to promote advanced biofuels. In EU-28, Renewable Energy Directive (RED) includes the "20/20/20" mandatory goals for 2020, one of which is a 20% share for renewable energy, and a 10% blending target for transport biofuels. The ILUC Directive of 2015 calls for a gradual reduction in the share of food-based biofuels in transport fuel, and a nonbinding national target for nonfood based (advanced) biofuels. The European Commission published RED II in 2016 as a legislative proposal for the period 2021–30. RED II calls for the European Union to produce at least 27% of its energy from renewable sources by 2030; to set a cap on conventional biofuels starting at 7% in 2021 and decreasing to 3.8% in 2030; and to support advanced biofuels with a minimum share of 1.5% in 2021 to 6.8% by 2030. Furthermore, RED II extends the existing biomass sustainability criteria for biofuels (REN21, 2017; GAIN Reports, 2018). The United States released 2017 blending mandates under its Renewable Fuel Standard, including 1.2 billion liters (311 million gallons) of cellulosic biofuels (which must achieve at least 60% lifecycle GHG emissions reductions) out of 4.28 billion gallons of advanced biofuels (which must achieve at least 50% lifecycle GHG emissions reductions) (REN21, 2017; EPA, 2017). Funding and grants were announced for advanced biofuels in Australia and the United States in 2016 (REN21, 2017). There is still a long way to go before advanced biofuels can be financially feasible and constitute a critical portion of biofuels use in transportation, but efforts are being made to diversify the biofuels value chain.

6. Conclusions and the way forward

Although the biofuels sector has grown significantly in the last decade, its global share in both the renewable energy and the traditional energy sectors remains low. Other forms of renewable energy that are gaining ground to tackle environmental concerns such as climate change and the use of biofuels in transportation forms only one piece of the puzzle. Despite its small role in renewable energy efforts, the first-generation biofuels sector utilizes a significant amount of agricultural feedstocks, thus having a large impact on the agricultural sector. This demand growth for agricultural feedstocks has further repercussions throughout upstream and downstream crop value chains, livestock and dairy value chains, fiber value chains, and agricultural input value chains, affecting all consumers and producers.

With the 2007–2008 and 2011 food price crises and ILUC from biofuels expansion, the short-term and long-term impacts of biofuels on these upstream and downstream value chains have received considerable interest from all economic agents along these value chains, as well as from policymakers. The ensuing food-fuel-fiber debate has called into question the viability of using agricultural feedstocks for transportation in the long-run.

Despite the fast growth of supply, there also remain obstacles to further biofuels expansion, such as (1) infrastructure bottlenecks affecting availability of FFVs or E85 fuel pumps, (2) demand bottlenecks leading to blending wall and low uptake of FFVs, and (3) partially waning political support. Even with these constraints, there have been sizeable consequences for food and fiber markets and consumers, although the estimates of these impacts from the empirical literature vary widely.

In the future, whether the competition of biofuels with food, feed, and fiber demand growth for feedstocks will intensify or not depends on:

(i) Market conditions

The relative competitiveness of biofuels with respect to traditional fossil fuels depends on energy prices. The recent decline in prices of fuels used in the transport sector has reduced the competitiveness of biofuels in blending and has thus limited the demand increase.

(ii) Policy support

Policy support takes various forms such as blending mandates and targets, tax exemptions or cuts, or grants. Despite the recent decline in energy prices, policy support for biofuels continues; especially, blending mandates have stabilized demand to a certain extent (WEO, 2016; REN21, 2017).

(iii) Nonagricultural feedstocks' feasibility

Advanced biofuels are seen as a way to ease the constraints of the food-fuel-fiber competition, but this can be materialized only if wastes and residues are utilized. If energy feedstocks or corn stover types of materials continue to be used, biofuels will continue to have an impact on the food, fiber, and feed markets. Advanced biofuels are not at a stage in which they can be construed as a feasible alternative. However, any effort to diversify the biofuels value chain is beneficial to ease the food-fuel-fiber competition.

(iv) Supply constraints

The food-fuel-fiber debate has mostly centered on natural resource constraints and how best to utilize these resources. Crop area reallocation and expansion was the main concern in this debate. However, land intensification is just as critical to increase supply, and efforts to aid yield growth globally would help alleviate this competition.

Finally, it is crucial to differentiate between the impact the biofuels sector has on markets in both the short and the long-run. These impacts depend on the degree of flexibility of the biofuels and energy policies, time required for agents to adapt to changes, and technical progress. Thus any discussion on this topic needs to differentiate between the long-term and the short-term effects of any policy change and design accordingly.

References

AFVD, 2017. U.S. Energy Information Administration, Renewable & Alternative Fuels, Alternative Fuel Vehicle Data. Available Online at https://www.eia.gov/renewable/afv/index.php.

Anderson, S., 2012. The demand for ethanol as a gasoline substitute. J. Environ. Econ. Manag. 63, 151–168.

Avalos, F., 2014. Do oil prices drive food prices? The tale of a structural break. J. Int. Money Financ. 42, 253–271.

Babcock, B., Pouliot, S., 2013. Price it and they will buy: How E85 can break the Blend Wall. Iowa State University. CARD Policy Briefs 13-PB 11, Available Online at https://www.card.iastate.edu/products/publications/pdf/13pb11.pdf.

Balcombe, K., Rapsomanikis, G., 2008. Bayesian estimation and selection of nonlinear vector error correction models: the case of the sugar-ethanol-oil nexus in Brazil. Am. J. Agric. Econ. 90 (3), 658–668.

Beach, R.H., Zhang, Y.W., McCarl, B.A., 2012. Modeling bioenergy, land use, and GHG emissions with FASOMGHG: model overview and analysis of storage cost implications. Clim. Change Econ. 3 (3), 1250012 (34 pages).

Biomass Research and Development Board (BRD), 2009. Increasing feedstock production for biofuels: Economic drivers, environmental implications, and the role of research. Technical report. Available Online at https://www.afdc.energy.gov/pdfs/increasing_feedstock_revised.pdf.

Brown, L., 1994. Facing food insecurity. In: Brown, et al. (Eds.), State of the World. W.W. Norton and Co., Inc, New York, pp. 177–197.

Calderini, D.F., Slafer, G.A., 1998. Changes in yield and yield stability in wheat during the 20th century. Field Crop Res. 57, 335–347.

Cassman, K.G., Doberman, A., Walters, D.T., Yang, Y., 2003. Meeting cereal demand while protecting natural resources and improving environmental quality. Annu. Rev. Environ. Resour. 28, 15–58.

Chen, X., Huang, H., Khanna, M., 2012. Land-use and greenhouse gas implications of biofuels: role of technology and policy. Clim. Change Econ. 3(3).

Condon, N., Klemick, H., Wolverton, A., 2015. Impacts of ethanol policy on corn prices: a review and meta-analysis of recent evidence. Food Policy 51, 63–73.

de Nicola, F., de Pace, P., Hernandez, M., 2016. Co-movement of major energy, agricultural, and food commodity price returns: a time-series assessment. Energy Econ. 57 (2016), 28–41.

Dodder, R.S., Kaplan, P.O., Elobeid, A., Tokgoz, S., Secchi, S., Kurkalova, L.A., 2015. Impact of energy prices and cellulosic biomass supply on agriculture, energy and the environment: an integrated modeling approach. Energy Econ. 51, 77–87.

Dumortier, J., Hayes, D., Carriquiry, M., Dong, F., Du, X., Elobeid, A., Fabiosa, J., Tokgoz, S., 2011. Sensitivity of carbon emission estimates from indirect land use change. Appl. Econ. Perspect. Policy 33 (3), 428–448.

Egli, D.B., 2008. Soybean yield trends from 1972 to 2003 in mid-western USA. Field Crop Res. 106, 53–59.

EIA, 2017. U.S. Energy Information Administration, State Energy Consumption Estimates—1960 Through 2014. Available Online at https://www.eia.gov/state/seds/seds-data-complete.php?sid=US#Consumption.

Elobeid, A., Tokgoz, S., Dodder, R., Johnson, T., Kaplan, O., Kurkalova, L., Secchi, S., 2013. Integration of agricultural and energy system models for biofuel assessment. Environ. Model. Softw. 48, 1–16.

EPA, 2016. U.S. Environmental Protection Agency, Lifecycle Greenhouse Gas Emissions for Select Pathways. Available Online at https://www.epa.gov/sites/production/files/2016-07/documents/select-ghg-results-table-v1.pdf.

EPA, 2017. U.S. Environmental Protection Agency, Final Renewable Fuel Standards for 2017, and the Biomass-Based Diesel Volume for 2018. Available Online at https://www.epa.gov/renewable-fuel-standard-program/final-renewable-fuel-standards-2017-and-biomass-based-diesel-volume.

ERS, 2011. Measuring the indirect land-use change associated with increased biofuel production; a review of modeling efforts. In: Report to Congress, February 2011.

ERS, 2018. U.S. Bioenergy Statistics Data. Available Online at https://www.ers.usda.gov/data-products/us-bioenergy-statistics/.

EU Official Journal RED-Directive 2009/28/EC. n.d. Available Online at http://eur-lex.europa.eu/legal-content/EN/TXT/HTML/?uri=CELEX:32009L0028&from=EN.

FAOSTAT, 2018. Database. Available Online at http://www.fao.org/faostat/en/#home.

Fargione, J., Hill, J., Tilman, D., Polasky, S., Hawthorne, P., 2008. Land clearing and the biofuel carbon debt. Science 319 (5867), 1235.

GAIN Reports, 2018. U.S. Foreign Agricultural Service GAIN Reports, various issues. Available Online at https://gain.fas.usda.gov/Pages/Default.aspx.

Hayes, D., Babcock, B., Fabiosa, J., Tokgoz, S., Elobeid, A., Yu, T., Dong, F., Hart, C., Chavez, E., Pan, S., Carriquiry, M., Dumortier, J., 2009. Biofuels: potential production capacity, effects on grain and live-stock sectors, and implications for food prices and consumers. J. Agric. Appl. Econ. 41 (2), 465–491.

Hochman, G., Zilberman, D., 2016. Corn ethanol and US biofuel policy ten years later: a systematic review and meta-analysis. In: Selected Presentation at Agricultural & Applied Economics Association Annual Meeting, Boston, MA, July 31–August 2, 2016.

IRENA, 2016. International Renewable Energy Agency Innovation Outlook Advanced Liquid Biofuels, Summary for Policy Makers. Available Online at http://www.irena.org/publications/2016/Oct/Innovation-Outlook-Advanced-Liquid-Biofuels.

Keeney, R., Hertel, T.E., 2009. The indirect land use impacts of United States biofuel policies: the importance of acreage, yield, and bilateral trade responses. Am. J. Agric. Econ. 91 (4), 895–909.

Khanna, M., Zilberman, D., 2012. Modeling the land-use and greenhouse-gas implications of biofuels. Clim. Change Econ. 3(3).

Laborde, D., Valin, H., 2012. Modeling land-use changes in a global CGE: assessing the EU biofuel mandates with the MIRAGE-BioF model. Clim. Change Econ. 3 (3), 1250017.

Laborde, D., Majeed, F., Tokgoz, S., Torero, M., 2016. Long-Term Drivers of Food and Nutrition Security. IFPRI Discussion Paper 01531, May.

Lucotte, Y., 2016. Co-movements between crude oil and food prices: a post-commodity boom perspective. Econ. Lett. 147, 142–147.

Menz, K.M., Pardey, P., 1983. Technology and U.S. corn yields: plateaus and price responsiveness. Am. J. Agric. Econ. 65 (3), 558–562.

National Research Council Report, 2011. National Research Council Committee on Economic and Environmental Impacts of Increasing Biofuels Production. 2011. Renewable Fuel Standard: Potential Economic and Environmental Effects of U.S. Biofuel Policy. National Academies of Sciences, Washington, DC.

OECD/FAO, 2017. OECD-FAO Agricultural Outlook. OECD Agriculture statistics (database) Available Online at. https://doi.org/10.1787/agr-data-en.

Oram, P.A., Hojjati, B., 1995. The growth potential of existing agricultural technology. In: Islam, N. (Ed.), Population and Food in the Early 21st Century: Meeting Future Food Demand of an Increasing World Population. Occasional Paper, International Food Policy Research Institute (IFPRI), Washington, DC, pp. 167–190.

Pingali, P.L., Hossain, M., Gerpacio, R.V., 1997. Asian Rice Bowls: The Returning Crisis. CAB International, Wallingford.

Poudel, B.N., Paudel, K.P., Timilsina, G., Zilberman, D., 2012. Providing numbers for a food versus fuel debate: an analysis of a future biofuel production scenario. Appl. Econ. Perspect. Policy 34 (4), 637–668.

Pouliot, S., Babcock, B., 2014. Impact of Ethanol Mandates on Fuel Prices when Ethanol and Gasoline are Imperfect Substitutes. Iowa State University. CARD WP 14-WP 551.

Reilly, J.M., Fuglie, K.O., 1998. Future yield growth in field crops: what evidence exists? Soil Tillage Res. 47, 275–290.

REN21, 2017. Renewables 2017 Global Status Report. Available Online at http://www.ren21.net/status-of-renewables/global-status-report/.

Searchinger, T., Heimlich, R., Houghton, R., Dong, F., Elobeid, A., Fabiosa, J.F., Tokgoz, S., Hayes, D.J., Yu, T., 2008. Factoring greenhouse gas emissions from land use change into biofuel calculations. Science, 1238–1240.

Serra, T., Zilberman, D., 2013. Biofuel-related price transmission literature: a review. Energy Econ. 37, 141–151.

Tokgoz, S., Laborde, D., 2014. Indirect land use change debate: what did we learn? Curr. Sustain. Renew. Energ. Rep. 1, 104–110.

Tyner, W.E., Petter, R., 2014. The potential for aviation biofuels—technical, economic, and policy analysis. Choices. 29 (1). first quarter 2014.

Tyner, W.E., Taheripour, F., Zhuang, Q., Birur, D., Baldos, U., 2010. Land Use Changes and Consequent CO2 Emissions Due to US Corn Ethanol Production: A Comprehensive Analysis. Department of Agricultural Economics, Purdue University. Final Report (revised) to Argonne National Laboratory. Available at: http://www.transportation.anl.gov/pdfs/MC/625.PDF.

Wisner, R., 2015. Estimated U.S. Dried Distillers Grains with Solubles (DDGS) Production & Use, Data. Available online at https://www.extension.iastate.edu/agdm/crops/outlook/dgsbalancesheet.pdf.

World Energy Outlook, 2016. International Energy Agency, World Energy Outlook 2016. OECD/IEA, Paris, France.

Zhang, W., Yu, E., Rozelle, S., Yang, J., Msangi, S., 2013. The impact of biofuel growth on agriculture: why is the range of estimates so wide? Food Policy 38, 227–239.

CHAPTER 6

Exploring the potential for riparian marginal lands to enhance ecosystem services and bioenergy production

Christine Costello*, Nasser Ayoub†
*University of Missouri, Columbia, MO, United States
†Helwan University, Cairo, Egypt

Contents

1. Introduction

Land, the resources it contains beneath the surface, and the biological activity that happens on it provide all critical human needs, including food, building supplies, and fuels. There is a limited amount of land available and an ever-increasing number of humans encroaching on it and negatively impacting many ecosystems and natural cycles. For many decades, humans have been searching for renewable energy, and one such option is energy generation from biomass. Given that food production is a primary need, there have been many efforts to identify ways to utilize land that is not suitable for food crop production (marginal land) to produce biomass for energy as well as to restore ecosystem functioning and services.

Roughly 38% of the ice-free terrestrial land surface is utilized for agricultural purposes to produce grains, oilseeds, and livestock to support the nutritional needs of the human population (UNDESA, 2012). Furthermore, human activity has resulted in the degradation of 60% of ecosystem services over the past 50 years (UNDESA, 2012). Therefore the

Biofuels, Bioenergy and Food Security
https://doi.org/10.1016/B978-0-12-803954-0.00006-1
101

utilization of land to produce bioenergy crops raises concerns over potential competition with land used to produce food crops (Valcu-Lisman et al., 2016; Liu et al., 2017; Mehmood et al., 2017). Relatedly, increased use of land for energy crops in a globalized economy could lead to deforestation and/or plowing of grasslands to accommodate food and biobased energy demands, further exacerbating ecological damage (Farrell et al., 2006; Searchinger et al., 2008). A proposed solution to these challenges is to cultivate cellulosic plant species, e.g., switchgrass, poplar, and miscanthus, on marginal lands. Cultivation of riparian marginal lands in particular could offer a few advantages such as improved water quality through nutrient, pesticide, and sediment attenuation from neighboring agricultural lands (Lee et al., 2003; Tomer et al., 2003). It could also have the potential to reduce lifecycle energy and greenhouse gas emissions (GHGs) through utilizing waterborne transportation, which uses much less energy and is less GHG-intensive per ton-mile transported than land- or air-based transportation.

There is no consistent agreement on the definition of marginal lands. Generally speaking, the term "marginal" is applied when it is not possible to profitably cultivate commodity crops on land. The underlying reasons for this lack of profitability arises from a number of physical reasons associated with the land or the location of the land. Physical reasons associated with the land itself may include, for example, poor soils, steep slopes, eroded soil, and high salinity (Goerndt and Mize, 2008; Cai et al., 2011; Hartman et al., 2011). This lack of a consistent definition and the dynamic nature of agricultural economics in a global market make it challenging to accurately estimate the amount of marginal land available globally, or even regionally. This chapter provides an overview of definitions for marginal land from a variety of authors across the globe in rural and urban landscapes.

Perennial, cellulosic species are the most promising options for planting on marginal lands both for bioenergy and restoration of ecosystem services. These plants have a high cellulose content, i.e., complex carbohydrates made up of over 3000 glucose units, and have been the focus of much research because biofuels derived from these plant species are expected to result in net reductions in GHGs compared to fossil-based fuels (Farrell et al., 2006; Field et al., 2007). Furthermore, many have posited that growing these cellulosic crops on degraded lands could achieve many additional ecological objectives, e.g., provide habitats for wildlife, improve soil health, sequester carbon, and avoid competition with food crops (Tilman et al., 2006; Williams et al., 2013). A number of these potential benefits are described herein.

It is generally understood that perennial, cellulosic plant species also require fewer fertilizer and pesticide inputs, slow runoff, and increased water infiltration (Mann and Tolbert, 2000; Parrish and Fike, 2005). Planting cellulosic perennial species within or at the edge of agricultural fields or in riparian areas can lead to interception and mitigation of nutrient and/or pesticide runoff to improve water quality within a watershed, while also providing biomass for energy (Lee et al., 2003; Tomer et al., 2003; Williams et al., 2013;

Asbjornsen et al., 2014). A holistic consideration of both energy production potential and ecosystem restoration potential should be considered when evaluating whether to cultivate marginal lands for energy.

Lifecycle assessment is a technique used to quantify the full environmental or economic costs of producing a product, process, or service over the entire supply chain (Fokaides and Christoforou, 2016; Hong et al., 2018). Due to the lower quality of marginal land and the potential for fragmented and remote land parcels, there are additional considerations beyond the quality of the land when evaluating the potential for these lands to produce bioenergy crops such that the life cycle energy and GHG emissions are less than existing fossil-based sources. For example, a hilly land topography or land with limited access may prevent using high-capacity equipment and consequently bioenergy crops production becomes infeasible (Ayoub et al., 2007). Low yields or long travel distances can result in net GHG increases and more energy expenditure than that recovered via the bio-based fuel and must also be considered when developing bioenergy projects on marginal lands. This is particularly important to consider given that a primary motivation for producing energy from biomass is to reduce lifecycle GHGs compared to fossil fuels and achieve a positive energy balance.

In this chapter the following topics are presented: a literature review of definitions used to characterize land as marginal; potential advantages and disadvantages of cultivating biomass on marginal lands for energy with emphasis on riparian areas; and a discussion of the estimated energy available through the cultivation of marginal lands in a variety of contexts.

2. Definitions of marginal lands

To avoid the cultivation of energy crops at the expense of food crops, a great deal of research has explored the potential for cultivating perennial, cellulosic crop species on marginal lands where traditional food crops are likely to fail (Mehmood et al., 2017). While yields on marginal lands can be lower than on prime cropland, many researchers have found that it is possible to successfully produce site-specific cultivars on marginal lands (Ahmad et al., 2015; Alexopoulou et al., 2017; Bandaru et al., 2015). An additional benefit of planting these crops on marginal lands is the potential for improving ecosystem functioning on these degraded lands (Bandaru et al., 2014). However, it seems that there is no one consistent definition of marginal lands between researchers and practitioners in the field. In this section, an overview of the many definitions that researchers have used to categorize land as marginal in the literature spanning 1979–2019 are discussed. Some of these definitions are summarized in Table 6.1.

As demonstrated by Table 6.1, the definitions of marginal lands mean different things in different contexts depending on the research area and study targets. Furthermore, in marginal definitions, the meaning of marginality is always given first followed by a reason

Table 6.1 Selected definitions from different research disciplines

Marginal land definition	Research category	Location	Year	References
Land at the margins of production, where potential returns are at a breakeven point with production costs that are defined by the local economic context, which vary regionally and temporally. This is due to degradation of soil quality coupled with the opening of more productive croplands to the west.	Second-generation bioenergy feedstock (switchgrass [*Panicum virgatum*], miscanthus [*Miscanthus × giganteus*], and willow [*Salix* spp.]).	Northeast USA	2014	Stoof et al. (2014)
Land that has previously been used for agriculture then abandoned, perhaps for social or economic reasons, could not be used for crop production due to specific environmental constraints, or where land is in a degraded state.	Ecosystem services impact of second-generation bioenergy crop production.	USA	2015	Holland et al. (2015)
Lands that are not currently under cultivation due to loss of fertility and high cost of chemical fertilizers.	Agriculture expansion in marginal land.	Mexico	1983	Doolittle (1983)
Rocky lands incapable of cultivation with little topsoil, semidesert areas, and rough or sloping lands.	Tree crops planting as a source of food in marginal lands.	USA	1979	MacDaniels and Lieberman (1979)
Land presently not used for agricultural production, residential purposes, and other social uses.	Second-generation bioenergy feedstock.	China	2016	Xue et al. (2016)
Lands not worth cultivating with agrofood crops because of a number of biophysical or economic constraints like low soil quality, water and salinity stress, extreme climate conditions, soil and terrain handicaps, long distance to the market, or state intervention, among others.	Bioenergy feedstock, giant reed.	Spain	2015	Sánchez and Fernández (2017)

Definition	Topic	Country	Year	Reference
Land that is heavily eroded.	Second-generation bioenergy crop production, *Miscanthus lutarioriparius*.	China	2014	Mi et al. (2014)
Lands where cost-effective production is not possible under given environmental conditions, cultivation techniques, agricultural management, as well as other economic and legal conditions, including lands such as idle or fallow cropland, abandoned or degraded cropland, and abandoned pastureland. Marginal lands are typically characterized by low productivity and reduced economic return or by severe limitations for agricultural use. They are generally fragile and at high environmental risk.	Concept, assessment, and management.	USA	2013	Kang et al. (2013a,b)
Lands that are not used for conventional crops.	Potential biodiversity conflicts.	South Africa	2015	Blanchard et al. (2015)
Marginal, often contaminated, sites exist in large areas across the world as a result of historic activities such as industry, transportation, and mineral extraction.	Rehabilitation in the context of phytoremediation.	Europe	2014	Andersson-Sköld et al. (2014)
Lands where cost-effective production is not possible under given environmental conditions, cultivation techniques, agricultural management, as well as other economic and legal conditions, including lands such as idle or fallow cropland, abandoned or degraded cropland, and abandoned pastureland. It normally has lower agricultural productivity, due to its less fertile soils and often less favorable water, climate, and possibly other environmental conditions.	Ecosystem modeling and bioenergy crops.	USA	2015	Qin et al. (2015)

Continued

Table 6.1 Selected definitions from different research disciplines—cont'd

Marginal land definition	Research category	Location	Year	References
Land on which cost-effective food and feed production is not possible under given site conditions and cultivation techniques. Some types of lands were defined such as degraded land, wasteland, abandoned agricultural land, etc.	Bioenergy production.	Netherlands	2011	Wicke (2011)
Nonprofitable land for food crops due to low productivity.	Bioenergy production potential.	Canada	2012	Tingting et al. (2012)
Vacant lands and underutilized public and private areas with adequate soil quality and sunlight.	Bioenergy production potential.	USA	2015	Saha and Eckelman (2015)
Lands typically characterized by low productivity and reduced economic return or by severe constraints for agricultural cultivation. They are generally fragile, and their use is environmentally risky. They have four categories: physically marginal land, biologically marginal land, environmentally and ecologically marginal land, and economically marginal land.	Land use planning.	USA	2013	Kang et al. (2013a,b)
Land that is not suitable for food crop production, low in nutrients, or a contaminated site.	Soil remediation.	Italy	2015	Accardi et al. (2015)
Lands that are unsuitable for crop production, but ideal for the growth of energy plants with high stress resistance. These lands include barren mountains, barren lands, and alkaline lands.	Bioenergy production potential.	China	2012	Lu et al. (2012)

Description	Topic	Country	Year	Reference
Lands poorly suited for food crops because of low productivity due to inherent edaphic or climatic limitations or because they are located in areas that are vulnerable to erosion or other environmental risks when cultivated.	Biofuel production (review).	USA	2014	Lewis and Kelly (2014)
Lands with fewer economic returns due to some biophysical constraints.	Value characterizations.	Nigeria	2015	Aliu (2015)
Lands of low average temperature, which are poorly suited to field crops because of low site productivity or unfavorable terrain characteristics.	Bioenergy production potential.	Germany	2016	Schweier et al. (2016)
The designation of land as marginal frequently results from high erodibility or past cropping history involving significant erosion losses.	Biomass crops.	USA/Canada	1998	Kort et al. (1998)
Marginal lands are noncroplands and/or lands not suitable for long-term food crop production as a result of high erodibility (K factor >0.4) and/or prone to flood (frequently flooded).	Bioenergy production potential.	USA	2019	Ayoub et al. (in press)

for being marginal, which in most cases is a physical or economic reason. The following subsections place these definitions into a more structured context from different viewpoints. In Section 2.1 a general analysis of marginal lands definitions and their different categories is given and in Section 2.2 marginal lands in the context of urban lands and their benefits are discussed.

2.1 Marginal lands definitions categories

The literature review is organized broadly into three main categories identified in the reviewed literature. These categories, as shown in Fig. 6.1, are: Economic or physical reasons, Production status, Land cover type.

Economic or physical reasons

In the economic reasons category, the term "marginal" is usually applied to land when production in a given time is unable to return a profit or the land has a low potential for profit (Hartman et al., 2011; Skevas et al., 2014; Stoof et al., 2014; Sánchez and Fernández, 2017). Given the fluctuating nature of markets and commodity prices, land

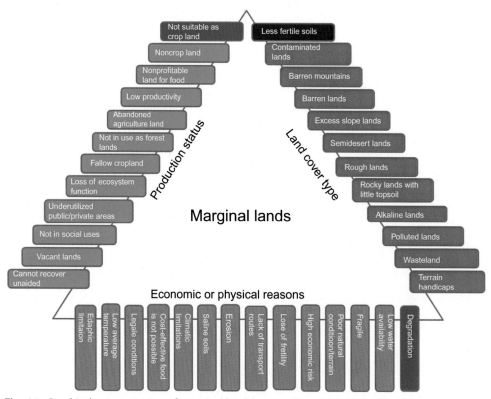

Fig. 6.1 Graphical representation of marginal land key words in the reviewed literature.

may be marginal in this sense in one year and not in another. A commonly applied definition for marginal land is land that is not suitable for conventional agricultural crops, i.e., commodity crops like corn, soy, and wheat (Gelfand et al., 2013). Where cost-effective production of food crops is not possible or nonprofitable on land it is considered marginal by many researchers no matter the underlying reason that causes the marginality (Wicke, 2011; Tingting et al., 2012; Aliu, 2015; Qin et al., 2015). In addition to these physical characteristics that prevent successful crop production, Kang et al. (2013a,b) and Sánchez and Fernández (2017) defined them as lands where cost-effective production is not possible under given environmental or climatic conditions.

Physical reasons for land being thought of as marginal are many and sometimes multiple reasons are given. Goerndt and Mize defined marginal land as "bottomlands with excess moisture, excessively dry lands with ridge tops or high slopes, and lands that have low-quality soils for farming" (Goerndt and Mize, 2008). Cai et al. considered marginal land as saline land, steep hillsides, and idle land (Cai et al., 2011). Others defined marginal lands as those lands with limited potential for food production due to high erodibility (Hartman et al., 2011), low soil fertility, steep slope, or poor drainage (MacDaniels and Lieberman, 1979; Andersson-Sköld et al., 2014; Sánchez and Fernández, 2017). Stoof et al. cited degradation of soil quality coupled with the cultivation of croplands in the western United States at the start of the 19th century as the beginning of marginality (Stoof et al., 2014). Researchers have also defined marginal lands as poorly drained fields that do not favor food crop production, lands prone to hazard events, and flood-prone land (Maantay and Maroko, 2009; Viator et al., 2012). Mi et al. gave a very general definition of marginal land that is heavily eroded (Mi et al., 2014). Kort et al. designated lands as marginal due to high erodibility or past cropping history involving significant erosion losses (Kort et al., 1998). Other reasons are typically linked to a physical condition or conditions of the land, including poor soil health, steep slopes, extensive erosion, salinity, etc. For instance, Gelfand et al. defined marginal lands as being poorly suited to field crops because of their low crop productivity due to inherent edaphic conditions, climatic limitations, vulnerable location, erosion potential, or other environmental risks when cultivated (Gelfand et al., 2013).

Production status

In this category, land that was once suitable for crop production is now marginal due to mismanagement of land over time. For example, Doolittle defined lands that are not currently under cultivation as marginal and gave loss of fertility and high cost of chemical fertilizers as the reason for vulnerability (Doolittle, 1983). Field et al. similarly defined marginal lands as lands formerly utilized for crop production but that have been abandoned, presumably due to degraded conditions (Field et al., 2007). Holland et al. defined land that has previously been used for agriculture and then abandoned, perhaps for social or economic reasons, as not suitable for crop production due to specific environmental constraints, or where land is in a degraded state (Holland et al., 2015). They also gave another definition of land not presently used for agricultural production, residential purposes, and other social uses (Xue et al., 2016). In the same class, lands were defined as abandoned agricultural lands, lands that are not used for conventional crops, lands with lower agricultural productivity, vacant lands, and underutilized public and private areas (Wicke, 2011; Blanchard et al., 2015; Qin et al., 2015; Saha and Eckelman, 2015). Kang et al. defined lands as typically characterized by low productivity and reduced economic return or by severe constraints for agricultural cultivation (Kang et al., 2013a,b).

2.2 Marginal lands in urban contexts

While the focus on cultivation of marginal land to produce bioenergy crops tends to be on rural areas, numerous researchers have focused on urban or peri-urban marginal lands to produce biomass for energy. The land types identified as marginal, the reasons for such classification, and recommended plant species often differ from rural and agricultural areas. For instance, the cause of marginality in urban and peri-urban environments is more likely to be due to contamination associated with industrial activity and the land parcels tend to be smaller. For example, Andersson et al. defined large areas of contaminated sites across the world as marginal lands as a result of historic activities such as industry, transportation, and mineral extraction (Andersson-Sköld et al., 2014). Likewise, Accardi et al. defined marginal lands as lands not suitable for food crop production, low in nutrients, or contaminated sites (Accardi et al., 2015). Lands that are marginal due to contamination in urban areas present both an opportunity for the added benefit of phytoremediation and a challenge due to the potential toxicity of existing contamination, which may require remediation prior to cultivation.

While land and biomass production potential tend to be smaller in these urban contexts, researchers have demonstrated that contributions from these lands can be considerable, offer the additional benefit of contaminant remediation, and may also offer social benefits. Saha and Eckleman quantified the potential marginal land available within a 30-mile radius of Boston, Massachusetts, USA (Saha and Eckelman, 2018). They identified potential parcels greater than $93\,m^2$ using geographic information system layers for land use descriptors and land cover types, impervious cover status, water and conservation land, and soil quality, including slope, and identified a total of $712\,km^2$. These researchers found that by planting short-rotation hybrid poplar on this land area, biomass could potentially fulfill all of the core city of Boston's heating needs (Saha and Eckelman, 2018). French et al. found that *Salix*, *Populus*, and *Alnus* could be economically harvested and could reduce levels of cadmium and zinc to acceptable levels over a 25–30-year period in northwestern England (French et al., 2006). In the city of Pittsburgh, Pennsylvania, USA, Niblick et al. used geographic information systems, National Resource Conservation Service soil classifications, and additional data from a nonprofit organization growing sunflowers on parcels in the city to improve identification of marginal land in the area (Niblick et al., 2013). They found that on approximately 3500 ha of land up to 129,000 L of sunflower-based biodiesel could be produced, and also highlighted the social benefits of community engagement and urban revitalization (Niblick et al., 2013).

3. Riparian buffer zones

Riparian buffers are vegetated areas next to waterways, which provide wildlife habitat, sediment, and nutrient sinks to improve water quality, reduce soil erosion, and potentially produce lignocellulosic feedstock for biofuels (Williams et al., 2015; Gu and Wylie, 2018). In addition, converting riparian buffers in agricultural areas from annual row crops to short-rotation cellulosic crops grown for biofuel production can offer economic motivation for buffer adoption among farmers (Rosa et al., 2017). However, great attention should be given to weed management when they are used for producing monoculture lignocellulosic feedstocks (Williams et al., 2015). In addition to the economic potential of planting energy crops in riparian zones there may also be great environmental benefit if the zones are unfertilized. The significance of energy willow as nutrient interception and retention on the environmental balance of a 50-m riparian buffer was tested in Sweden. The planted filter willow was found to achieve net global warming potential (GWP) and eutrophication potential (EP) savings of up to $11.9\,Mg\,CO_2e$ and $47\,kg\,PO_4e\,ha^{-1}\,year^{-1}$, respectively, compared with a GWP saving of $14.8\,Mg\,CO_2e\,ha^{-1}\,year^{-1}$ and an EP increase of $7\,kg\,PO_4e\,ha^{-1}\,year^{-1}$ for fertilized willow (Styles et al., 2016).

The restoration of degraded riparian zones can provide many ecosystem services. Fortier et al. evaluated ecosystem service enhancement potential after conversing the nonmanaged herbaceous buffers with hybrid poplar (*Populus* spp.) replacement buffers in southern Québec. This placement of riparian zones alongside farm streams could lead to the production of 5280–76,151 tons of poplar with the potential of displacing 2–29 million L of fuel oil (Fortier et al., 2016). Also, Gu studied the replacement of cropland waterway buffers with high switchgrass productivity potential in the eastern Great Plains (EGP) and concluded that about $16,342\,km^2$ of waterway buffers in the EGP are potentially suitable for switchgrass development with an estimated 15 million MT of switchgrass production every year (Gu and Wylie, 2018).

3.1 A case study: Bioenergy cultivation in the Missouri/Mississippi River Corridor

In the United States, biofuel production from planting energy crops on marginal lands is thought to be a very promising renewable energy source (Wicke, 2011; Stoof et al., 2014; Zhao et al., 2014). However, to achieve this goal, the high logistics cost of biomass resources must be solved (Ayoub et al., 2007, 2009). One of the proposed solutions is to produce energy crops along river corridors to maximize transportation via waterways, e.g., using barges to transfer biomass to biorefineries. Planting biomass species in riparian areas can provide ecosystem services by reducing nutrient and pesticide loading to the waterway. Furthermore, maximizing water-based transportation can considerably reduce lifecycle energy and GHGs since water-based transportation results in lower fuel consumption per ton-mile traveled than other modes of transport. One of the projects that is trying to realize this potential is the Mississippi/Missouri River Advanced Biomass/Biofuel Consortium (MRABC) collaborative project between more than 40 academic institutions and agricultural and energy companies working to turn the Missouri/Mississippi River Corridor (MMRC) into a provider of clean energy for the United States and economic opportunity for Missouri (Jose, 2015). One of the aims of the MRABC project is to demonstrate the capacity of the Midwest to be a significant contributor to biofuel production in the United States without encroaching on land dedicated to food crops. Another aim is to investigate how to transport and refine the biomass profitably and with added ecosystem benefits via cultivating marginal lands along the rivers. To show the potential of this idea, the current distribution of barge ports and biorefineries around MMRC, within 100 miles of the Mississippi/Missouri Rivers centerline, is shown in Figs. 6.2 and 6.3, respectively. Such an affordable way to transport energy crops is essential to making biomass production feasible for growers and bioenergy producers. This win–win situation can create an established economically stable market that can take biofuels to another stage, for example, the creation of regional biofuel processing and refining plants along the river. These facilities close to harvest sites will process the biomass into easier-to-ship bioenergy products such as pellets and fuels that will allow economic methods of shipping, e.g., trucks, throughout the United States or through the Port of New Orleans for export.

In a recent publication, the area of marginal lands available in the MMRC under a variety of scenarios was determined (Ayoub et al., in press). The scenarios were developed to include physical reasons for marginal land status, specifically lands that are classified as "frequently flooded" and/or lands with high erodibility factors defined by a factor K greater than 0.4, as defined by the Natural Resources Conservation Service's Soil Survey Geographic database (NRCS-SSURGO, 2017). To explore other policy and environmental considerations, scenarios were also developed to include or exclude lands that met the physical criteria for marginality, but that were also forests or Conservation Reserve Program (CRP) lands. Forested areas and CRP lands are typically not permissible for the use of bioenergy cultivation. While evidence generally supports that forested land should not be utilized for bioenergy cultivation, the potential for achieving the mission of conservation while also collecting biomass from CRP lands is less obvious; see earlier. The estimated range of land available across 24 scenarios ranged from 0.1 Mha in the most conservative case to 58.4 Mha in the case where marginality could be defined as either frequently flooded or highly erodible, and forest and CRP lands were considered

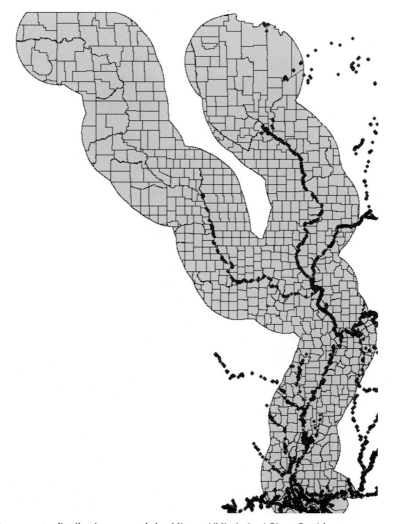

Fig. 6.2 Barge ports distribution around the Missouri/Mississippi River Corridor.

eligible (Ayoub et al., in press). The authors identified two scenarios as most likely to be achievable and produce considerable amounts of biomass. These scenarios included lands that were either frequently flooded or highly erodible, or erodible only and did not include forest or CRP lands, but that could have been classified as cropland, but were deemed unsuitable for crop production given the floodability or erodibility status. In these scenarios, 27.56 or 30.39 Mha were identified in the MMRC (Ayoub et al., in press). A range of observed yield data for switchgrass, willow, and poplar within the study region and ranges of observed energy conversion factors for ethanol, jet fuel, and

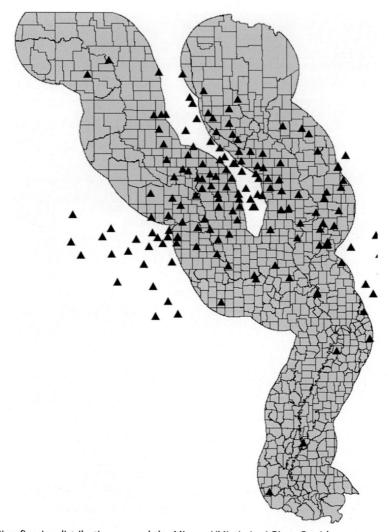

Fig. 6.3 Biorefineries distribution around the Missouri/Mississippi River Corridor.

electricity were then employed to estimate energy potential. The authors estimated that these two scenarios could provide on average between 297 and 365 million tons of biomass, or between 98 and 131 billion L of ethanol; 63 and 76 billion L of biojet fuel; or 469 and 513 TWh of electricity (Ayoub et al., in press).

4. Estimates of marginal land available across a variety of scales

Estimations of the amount of marginal land available vary considerably, which is not surprising given the lack of consistent definition and the dynamic nature of land use and economics. In this section we briefly describe a few efforts at global, national, and smaller, e.g., city-wide, scales on rural and urban lands (Table 6.2). One often utilized approach is to use geographic information systems to compare land use across time to identify lands that were previously cultivated and currently no longer in use. The logic of this is that if the land were cultivated at some point and were once capable of sustaining crops, but now, for either a physical or economic reason, are no longer able to do so, these lands might be restored due to the cultivation of deep-rooting, perennial species. At the global scale, Field et al. applied this logic by identifying land that was used as cropland at some point over the years 1700–2000 to calculate the difference between maximum cropland and pasture area over this time and the area used for these purposes in 2000 (Field et al., 2007). Next, they excluded what became urban areas (3%), abandoned pasture that became cropland (33%), and land that became forested (13%) for a final global estimate of 386 Mha (Field et al., 2007) (Table 6.2).

Zhuang et al. estimated the amount of marginal land available in China using a set of physical criteria, including: land use type, elevation, slope, precipitation, temperature,

Table 6.2 Selected estimates of marginal land available across spatial scales

Geographical area	Marginal land estimate (Mha)	Criteria for marginality	Source
Globe	386	Abandoned crop or pasture land.	Field et al. (2007)
China	130.34 (43.75 exploitable)	Land use type, elevation, slope, precipitation, temperature, soil type, alkalization, and effective depth. Excluding stable land providing ecosystem services.	Zhuang et al. (2011)
China— miscanthus-focused evaluation	171.64 (7.694 suitable for miscanthus cultivation)	Not suitable for agricultural production and suitable precipitation and winter temperatures to support miscanthus.	Xue et al. (2016)
Mississippi/ Missouri River Corridor	27.56–30.39 (most viable options)	High erodibility and flooding potential within 161 km of the Missouri/Mississippi River Corridor.	Ayoub et al. (in press)
Boston environs— Metropolitan Area Planning Council (MAPC)	0.071 (20% of total land area in MAPC)	Vacant land and underutilized land with adequate soil quality and sunlight.	Saha and Eckelman (2015) and Saha and Eckelman (2018)

soil type, alkalization, and effective depth, and then excluded land uses that currently supported shrubland and high and moderate coverage grasslands for ecological reasons (Zhuang et al., 2011). They then identified bioenergy plant species, i.e., *Helianthus tuberosus* L., *Cassava*, and three woody oil plants (*Pistacia chinensis, Jatropha curcas* L., *Vernicia fordii*) that would likely be suitable for growth given the characteristics of the land and climate in the locations identified (Zhuang et al., 2011). Without exclusion of some shrub and grasslands deemed beneficial to ecosystem services the total estimated marginal land available was estimated to be 130.34 Mha. With these exclusions the estimated area was reduced to 43.75 Mha, with an estimated potential of 133.85 million tons of biomass (Zhuang et al., 2011).

5. Advantages and disadvantages of growing energy crops on marginal lands

As indicated earlier, in addition to cultivating biomass for energy production, in many cases such cultivation can also rehabilitate ecosystems and improve water quality. Bioenergy cultivation, particularly mixed species of vegetation, can create habitats to increase biodiversity and restore ecological health, particularly on degraded or abandoned lands (Tilman et al., 2006; Pedroli et al., 2013; Williams et al., 2013; Blank et al., 2014). It should be noted that mixed species plantings for bioenergy can result in lower total biomass yield, which may be an acceptable tradeoff (Adler et al., 2009). In addition, the increased diversity of plant matter can lead to further complications during energy conversion, e.g., high silica content can lead to fouling during combustion. Nonetheless, these plant species offer a variety of ecosystem services, including carbon sequestration, soil health, species habitat, nutrient attenuation, improved water quality, and flood protection (Williams et al., 2013; Asbjornsen et al., 2014). Therefore lower yields may be a worthwhile tradeoff.

Cellulosic species planted on lands with low biomass productivity, e.g., formerly cultivated and abandoned lands, can result in carbon sequestration (Harden et al., 1999; Tilman et al., 2006). Perennial plants store carbon in their rooting systems as well as in the soil and thus can sequester carbon from the atmosphere; this contributes to the net GHG benefits of the biofuel. It is important to note that the carbon sequestration benefit is not realized if an intact, productive grassland or forest is replaced with cultivated biomass species because the initial carbon loss from cutting down and plowing the soil is very large (Tilman et al., 2006; Fargione et al., 2008).

Exposed soils are at risk of erosion, and agricultural activity is the primary cause of increased soil erosion. Soil erosion is a serious concern for agricultural productivity as well as healthy ecosystems (FAO and ITPS, 2015; Borrelli et al., 2017). Soil erosion can lead to lack of sufficient growth medium for productive crop cultivation, sedimentation of waterways, mobilization of nutrients, poor soil health, and microbiomes

(FAO and ITPS, 2015). Rates of soil erosion are highest on cultivated land, though improvements in management practices, e.g., winter cover crops, have been shown to reduce soil erosion rates in the United States and Brazil (FAO and ITPS, 2015). Expansion of agricultural activity onto marginal lands increases the risk of soil erosion (FAO and ITPS, 2015). When previously cultivated lands, which may have been of poorer quality to begin with, are abandoned, continued risk of soil erosion is present. By planting these areas with deep-rooting perennials, such as cellulosic plant species, soils can be stabilized to mitigate erosion as well as to rebuild soil organic carbon content, and ultimately to regenerate healthy soil conditions. Therefore in instances where land is classified as marginal, if it is possible to cultivate and harvest biomass for energy this could offer a win-win situation of ecosystem services restoration as well as bioenergy (Williams et al., 2013; Asbjornsen et al., 2014; Blank et al., 2014).

To ensure that a bioenergy project positively contributes to biodiversity and carbon sequestration requires careful consideration of how the land is currently occupied, as well as continued management for the duration of the project (Tilman et al., 2006; Pedroli et al., 2013). Pedroli et al. conducted an extensive analysis over a number of European countries to identify the conditions under which conversion of land from one type to primary cultivation of perennial biomass crops leads to positive outcomes (Pedroli et al., 2013). They concluded that while it is unlikely due to economics that land used for food and feed production would be converted to use for biomass, interspersing these lands with shelter belts, wildlife corridors, or buffer strips would likely increase biodiversity (Pedroli et al., 2013). Conversion of lands currently occupied by intensively managed permanent grasses or fodder crops to perennial biomass species can result in a variety of positive and negative outcomes (Pedroli et al., 2013). For example, if the existing fodder is managed with fertilizers and pesticides and only contains a limited number of plant species, then conversion to a mixed species, low-input perennial cropping system could improve habitats; however, if the existing grasses or fodder are already low input, conversion to a monocrop of bioenergy species could result in a decrease in habitat diversity and possibly an increase in chemical inputs (Pedroli et al., 2013). For lands that are currently low-intensity agriculture, such as agroforestry areas, long-term fallow lands, permanent grasslands, or grazed shrubland, conversion to cultivation of individual biomass crops could decrease the variety of habitats or lead to soil erosion or even nutrient runoff (Pedroli et al., 2013). However, if these lands are under threat of being abandoned, selective removal of biomass using extensive management techniques could maintain habitat richness while gathering some biomass (Pedroli et al., 2013). In conclusion, it is important to consider what the land is currently supporting, whether it exhibits qualities of marginality or not, when determining the pros and cons of converting land to primarily support bioenergy species.

Marginal land may be lacking in species diversity, and in many cases the introduction of a variety of plant species that are able to stabilize and restore soils can create lasting habitats for a variety of other organisms. For example, Blank et al. explored the potential for three potential bioenergy grassland fields (grass monocultures, grass-dominated fields, and forb-dominated fields) to provide habitats for birds as well as produce biomass for energy uses (Blank et al., 2014). While there are tradeoffs associated with harvesting time to avoid disturbing birds at critical periods, which can lead to lower yield, the authors found that it was possible to produce considerable quantities of bioenergy feedstocks while also increasing bird habitats, particularly the forb-dominated fields (Blank et al., 2014). In a subsequent analysis, Blank et al. explored the significance of spatial arrangements of bioenergy plantings and found that clustering grasslands had the strongest benefit for bird habitats, as opposed to strategic citing for proximity to conversion facilities or random placement, but again emphasized the potential for cobenefits (Blank et al., 2016). Thus a key consideration in proceeding with bioenergy projects that utilize marginal lands and also aspire to increase habitat for wildlife is stakeholder engagement and collaborative decision-making (Williams et al., 2013). The priorities vary considerably between an entrepreneur attempting to profit from bioenergy, a policymaker tasked with decreasing GHGs associated with energy systems, a conservation ecologist, and wildlife enthusiasts, thus building projects with all parties is critical to achieving mutual satisfaction (Williams et al., 2013).

The issue of invasive species presents both an opportunity and a risk in relation to bioenergy production potential. Invasive species are nonnative to the region and if they have an advantage over native species in terms of reproduction or growth they can eventually dominate the landscape to the detriment of existing native species. Due to the intricate relationships developed over time in ecosystems, this sudden domination by a new species can be detrimental to healthy ecosystem functioning.

The present opportunity is to utilize biomass obtained through ecosystem restoration projects that remove large quantities of invasive species (Nackley et al., 2013; VanMeerbeek et al., 2015). When evaluating the potential feasibility of such projects, considerations include any biomass preparation required beyond that required for ecosystem restoration, density of the biomass in the area, distance to the bioenergy processing operation, and the chemical composition of the biomass in relation to the energy conversion technology. Nackley et al. found that removal of Russian olive biomass from areas in Washington state, where the species made up over 80% of the riparian canopy, could be economically viable when transportation costs were under $9.5/MT of dry biomass combusted in a biomass-fired boiler (Nackley et al., 2013). Overall, given the extent of the invasion of biomass species in the region, extrapolated to the western United States, Nackley et al. estimated that 5.7–39 million Mg year^{-1} of salt cedar and 5.8–39.7 million Mg year^{-1} of Russian olive could be harvested (Nackley et al., 2013). Numerous other researchers have documented the potential of removing biomass during conservation management activities for energy production, including anaerobic digestion of invasive annual and perennial herbaceous species (VanMeerbeek et al., 2015) and anaerobic codigestion of swine manure and water hyacinth (Lu et al., 2010). Thirteen invasive species were evaluated and nonwoody species were found suitable for electricity generation through combustion, while high-moisture species were found to be more appropriate for biochemical conversion pathways (Melane, 2016).

On the other hand, dedicated energy crops may have the potential to become invasive species and thus it is important to evaluate this potential risk when developing bioenergy projects (Raghu et al., 2006; Barney and DiTomaso, 2008, 2011). Many bioenergy crops are not native to the areas available for cultivation and it is often unknown how large-scale introduction of a new plant species may impact the existing ecosystem (Raghu et al., 2006). Furthermore, traits desirable for the cultivation of bioenergy crops on marginal lands that occur naturally or through breeding or genetic engineering, such as pest resistance, ability to thrive in poor growing conditions, minimal external input requirements, and potential for achieving high yields, may also increase the potential for the bioenergy species to overtake existing vegetation and become invasive (Barney and DiTomaso, 2008, 2011). The potential for a plant species to become invasive is also highly dependent on the geography and conditions in which it is planted and thus may have varying degrees of risk across locations.

To explore this question, Barney and DiTomaso applied a Weed Risk Assessment protocol to giant reed, switchgrass, and miscanthus in the United States and found that switchgrass could pose a risk in California, reed canary grass presented a risk in Florida, while miscanthus posed little threat (Barney and DiTomaso, 2008). In a later work, Barney and DiTomaso used a mathematical model (CLIMEX) to evaluate climate niches for grass, herbaceous, and woody bioenergy crops as well as invasive species to more deeply assess whether a bioenergy crop had significant risk of expanding invasively beyond areas in which it was planted (Barney and DiTomaso, 2011). This work showed that bioenergy crops have a potential to become invasive given broad climatic tolerance, but that the Weed Risk Assessment approach overestimated the potential compared to the modeling approach (Barney and DiTomaso, 2011). In summary, the introduction of a bioenergy crop could lead to unintended spread, or invasive status, and should be considered in each new project.

In addition to ecosystem considerations, there are economic, logistic, engineering, and political considerations when planning bioenergy projects on marginal lands. Many of the physical reasons that result in land being classified as marginal may also make harvesting, collecting, and transporting biomass to a bioenergy facility difficult, and thus more expensive (Ayoub et al., 2007). For example, steep slopes, poor accessibility, small and dispersed parcels, or low-quality soils may result in yields or transportation logistics that are not economically viable. It is important to recall that, while ecosystem services may be considerably restored via these projects, they are largely unvalued in our economic system. Thus the predominant determining factors in developing a bioenergy project is whether or not energy can be profitably produced and whether the biomass feedstock can be made consistent, in terms of both bulk quantity and chemical composition (Gold and Seuring, 2011). In addition to adequate yields without interfering with food crops, the supply chain, i.e., collection/harvesting, storage, transportation, pretreatment techniques, and the ultimate energy conversion technology, must all be systematically evaluated as a whole to ensure net GHG, energy, and ecosystem benefits (Gold and Seuring, 2011).

6. Conclusion

This chapter provided a review of marginal lands literature, providing a rich exploration of the many and variable reasons why land has been classified as such across many geographies and contexts. Cultivation in riparian corridors may offer unique ecosystem services as well as reduce lifecycle GHG and energy use over the lifecycle of the bio-based energy products generated from the cellulosic materials harvested. Many of the challenges and opportunities associated with cultivating bioenergy crops on marginal land and subsequently converting them into usable energy were presented.

The scale of analysis often influences the definition of marginal land applied. For example, when attempting to estimate national or global availability of marginal lands, an analyst is constrained to the quality of the datasets available over these large geographies. The unit of analysis in large-scale studies is often quite coarse and on-the-ground details may not be accounted for. Nonetheless, these estimates are critical for long-term strategic planning as humans strive to meet their food, energy, shelter, and healthcare needs. Smaller scales of analysis are able to highlight the extreme complexity in not only identifying marginal land, but also negotiating among all relevant stakeholders to ensure that the cultivation of bioenergy crops on the land will deliver a fuel that contributes fewer GHGs than fossil fuels and, hopefully also, restores ecosystem services to the degraded land area. Continued conversation across stakeholders, valuation of ecosystem services, and supportive policy are needed to realize the full potential of bioenergy from marginal lands.

References

Accardi, D.S., et al., 2015. From soil remediation to biofuel: process simulation of bioethanol production from Arundo donax. Chem. Eng. Trans. 43, 2167–2172.

Adler, P.R., et al., 2009. Plant species composition and biofuel yields of conservation grasslands. Ecol. Appl. 19 (8), 2202–2209.

Ahmad, M., et al., 2015. Optimization of biodiesel production from *Carthamus tinctorius* L. CV.Thori 78: a novel cultivar of safflower crop. Int. J. Green Energy 12 (5), 447–452.

Alexopoulou, E., et al., 2017. Long-term studies on switchgrass grown on a marginal area in Greece under different varieties and nitrogen fertilization rates. Ind. Crop Prod. 107, 446–452.

Aliu, I.R., 2015. Marginal land use and value characterizations in Lagos: untangling the drivers and implications for sustainability. Environ. Dev. Sustain., 1–20.

Andersson-Sköld, Y., et al., 2014. Developing and validating a practical decision support tool (DST) for biomass selection on marginal land. J. Environ. Manage. 145, 113–121.

Asbjornsen, H., et al., 2014. Targeting perennial vegetation in agricultural landscapes for enhancing ecosystem services. Renew. Agric. Food Syst. 29 (2), 101–125.

Ayoub, N., et al., 2007. Two levels decision system for efficient planning and implementation of bioenergy production. Energy Convers. Manage. 48 (3), 709–723.

Ayoub, N., et al., 2009. Evolutionary algorithms approach for integrated bioenergy supply chains optimization. Energy Convers. Manage. 50 (12), 2944–2955.

Ayoub, N., et al., in press. Systematic application of a quantitative definition of marginal lands in estimating biomass energy potential in the Missouri/Mississippi River Corrido. *Biofuels*, https://doi.org/10.1080/17597269.2018.1554945.

Bandaru, V., et al., 2014. Soil carbon change and net energy associated with biofuel production on marginal lands: a regional modeling perspective. J. Environ. Qual. 42 (6), 1802–1814.

Bandaru, V., Parker, N.C., Hart, Q., Jenner, M., Yeo, B.-L., Crawford, J., Li, Y., Tittmann, P., Rogers, L., Kaffka, S., Jenkins, B., 2015. Economic sustainability modeling provides decision support for assessing hybrid poplar-based biofuel development in California. Calif. Agric. 69 (3), 171–176.

Barney, J.N., DiTomaso, J.M., 2008. Nonnative species and bioenergy: are we cultivating the next invader? BioScience 58 (1), 64–70.

Barney, J.N., DiTomaso, J.M., 2011. Global climate niche estimates for bioenergy crops and invasive species of agronomic origin: potential problems and opportunities. PLoS One 6 (3).

Blanchard, R., et al., 2015. Anticipating potential biodiversity conflicts for future biofuel crops in South Africa: incorporating spatial filters with species distribution models. GCB Bioenergy 7 (2), 273–287.

Blank, P.J., et al., 2014. Bird communities and biomass yields in potential bioenergy grasslands. PLoS One 9 (10).

Blank, P.J., et al., 2016. Alternative scenarios of bioenergy crop production in an agricultural landscape and implications for bird communities. Ecol. Appl. 26 (1), 42–54.

Borrelli, P., et al., 2017. As assessment of the global impact of 21st century land use change on soil erosion. Nat. Commun. 8, 13.

Cai, X., et al., 2011. Land availability for biofuel production. Environ. Sci. Technol. 45 (1), 334–339.

Doolittle, W.E., 1983. Agricultural expansion in a marginal area of Mexico. Geogr. Rev. 73 (3), 301–313.

FAO and ITPS, 2015. Status of the World's Soil Resources (SWSR)—Main Report. Food and Agriculture Organization of the United Nations and Intergovernmental Technical Panel on Soils.

Fargione, J., et al., 2008. Land clearing and the biofuel carbon debt. Science 319 (5867), 1235–1238.

Farrell, A.E., et al., 2006. Ethanol can contribute to energy and environmental goals. Science 311 (5760), 506–508.

Field, C.B., et al., 2007. Biomass energy: the scale of the potential resource. Trends Ecol. Evol. 23 (2), 65–72.

Fokaides, P.A., Christoforou, E., 2016. Life cycle sustainability assessment of biofuels. In: Luque, R., Lin, C.S.K., Wilson, K., Clark, J. (Eds.), Handbook of Biofuels Production, second ed. Woodhead Publishing, pp. 41–60 (Chapter 3).

Fortier, J., et al., 2016. Potential for hybrid poplar riparian buffers to provide ecosystem services in three watersheds with contrasting agricultural land use. Forests 7 (2), 37.

French, C.J., et al., 2006. Woody biomass phytoremediation of contaminated brownfield land. Environ. Pollut. 141 (3), 387–395.

Gelfand, I., et al., 2013. Sustainable bioenergy production from marginal lands in the US Midwest. Nature 493 (7433), 514–517.

Goerndt, M.E., Mize, C., 2008. Short-rotation woody biomass as a crop on marginal lands in Iowa. North. J. Appl. For. 25 (2), 82–86.

Gold, S., Seuring, S., 2011. Supply chain and logistics issues of bio-energy production. J. Clean. Prod. 19, 32–42.

Gu, Y., Wylie, B.K., 2018. Mapping cropland waterway buffers for switchgrass development in the eastern Great Plains, USA. Glob. Change Biol. Bioenergy 10 (6), 415–424.

Harden, J., et al., 1999. Dynamic replacement and loss of soil carbon on eroding cropland. Global Biogeochem. Cycles 13 (4), 885–901.

Hartman, J.C., et al., 2011. Potential ecological impacts of switchgrass (*Panicum virgatum* L.) biofuel cultivation in the Central Great Plains, USA. Biomass Bioenergy 35 (8), 3415–3421.

Holland, R.A., et al., 2015. A synthesis of the ecosystem services impact of second generation bioenergy crop production. Renew. Sustain. Energy Rev. 46, 30–40.

Hong, J., et al., 2018. Life-cycle environmental and economic assessment of medical waste treatment. J. Clean. Prod. 174, 65–73.

Jose, S., 2015. Biomass/Biofuel Corridor Along the Mississippi/Missouri River. Retrieved Februery, 9, 2016, from: http://mizzouadvantage.missouri.edu/project/bio2cor-the-biomassbiofuel-corridor-along-the-mississippimissouri-river/.

Kang, S., et al., 2013a. Hierarchical marginal land assessment for land use planning. Land Use Policy 30 (1), 106–113.

Kang, S., et al., 2013b. Marginal lands: concept, assessment and management. J. Agric. Sci. 5, 5.

Kort, J., et al., 1998. A review of soil erosion potential associated with biomass crops. Biomass Bioenergy 14 (4), 351–359.

Lee, K.H., et al., 2003. Sediment and nutrient removal in an established multi-species riparian buffer. J. Soil Water Conserv. 58 (1), 1–7.

Lewis, S., Kelly, M., 2014. Mapping the potential for biofuel production on marginal lands: differences in definitions, data and models across scales. ISPRS Int. J. Geo-Inf. 3 (2), 430.

Liu, T., et al., 2017. Bioenergy production on marginal land in Canada: potential, economic feasibility, and greenhouse gas emissions impacts. Appl. Energy 205, 477–485.

Lu, J., et al., 2010. Integrating animal manure-based bioenergy production with invasive species control: a case study at Tongren Pig Farm in China. Biomass Bioenergy 34 (6), 821–827.

Lu, L., et al., 2012. Evaluating the marginal land resources suitable for developing *Pistacia chinensis*-based biodiesel in China. Energies 5 (7), 2165.

Maantay, J., Maroko, A., 2009. Mapping urban risk: flood hazards, race, & environmental justice in New York. Appl. Geogr. 29 (1), 111–124.

MacDaniels, L.H., Lieberman, A.S., 1979. Tree crops: a neglected source of food and forage from marginal lands. BioScience 29 (3), 173–175.

Mann, L., Tolbert, V., 2000. Soil sustainability in renewable biomass plantings. Ambio 29 (8), 492–498.

Mehmood, M.A., et al., 2017. Biomass production for bioenergy using marginal lands. Sustain. Prod. Consum. 9, 3–21.

Melane, M., 2016. Evaluation of the potential of non-woody invasive plant biomass for electricity generation. In: Forestry and Wood Science. Stellenbosch University, South Africa (Masters).

Mi, J., et al., 2014. Carbon sequestration by Miscanthus energy crops plantations in a broad range semi-arid marginal land in China. Sci. Total Environ. 496, 373–380.

Nackley, L.L., et al., 2013. Bioenergy that supports ecological restoration. Front. Ecol. Environ. 11 (10), 535–540.

Niblick, B., et al., 2013. Using geographic information systems to assess potential biofuel crop production on urban marginal lands. Appl. Energy 103, 234–242.

NRCS-SSURGO, 2017. Web Soil Survey. Retrieved January 21, 2019, from: https://websoilsurvey.sc.egov.usda.gov/App/HomePage.htm.

Parrish, D.J., Fike, J.H., 2005. The biology and agronomy of switchgrass for biofuels. Crit. Rev. Plant Sci. 24 (5–6), 423–459.

Pedroli, B., et al., 2013. Is energy cropping in Europe compatible with biodiversity?—opportunities and threats to biodiversity from land-based production of biomass for bioenergy purposes. Biomass Bioenergy 55, 73–86.

Qin, Z., et al., 2015. Bioenergy crop productivity and potential climate change mitigation from marginal lands in the United States: an ecosystem modeling perspective. GCB Bioenergy 7 (6), 1211–1221.

Raghu, S., et al., 2006. Adding biofuels to the invasive species fire? Science 313, 1742.

Rosa, D.J., et al., 2017. Water quality changes in a short-rotation woody crop riparian buffer. Biomass Bioenergy 107, 370–375 (Research paper).

Saha, M., Eckelman, M.J., 2015. Geospatial assessment of potential bioenergy crop production on urban marginal land. Appl. Energy 159, 540–547.

Saha, M., Eckelman, M.J., 2018. Geospatial assessment of regional scale bioenergy production potential on marginal and degraded land. Resour. Conserv. Recycl. 128, 90–97.

Sánchez, J., Fernández, J., 2017. Approach to the potential production of giant reed in surplus saline lands of Spain. GCB Bioenergy 9, 105–118

Schweier, J., et al., 2016. Selected environmental impacts of the technical production of wood chips from poplar short rotation coppice on marginal land. Biomass Bioenergy 85, 235–242.

Searchinger, T., et al., 2008. Use of U.S. croplands for biofuels increases greenhouse gasess through emissions from land-use change. Science 319 (5867), 1238–1240.

Skevas, T., et al., 2014. What type of landowner would supply marginal land for energy crops? Biomass Bioenergy 67, 252–259.

Stoof, C.R., et al., 2014. Untapped potential: opportunities and challenges for sustainable bioenergy production from marginal lands in the Northeast USA. Bioenergy Res. 8 (2), 482–501.

Styles, D., et al., 2016. Climate regulation, energy provisioning and water purification: quantifying ecosystem service delivery of bioenergy willow grown on riparian buffer zones using life cycle assessment. Ambio 45 (8), 872–884.

Tilman, D., et al., 2006. Carbon-negative biofuels from low-input high-diversity grassland biomass. Science 314 (5805), 1598–1600.

Tingting, L., et al., 2012. Bioenergy production potential on marginal land in Canada. In: 2012 First International Conference on Agro-Geoinformatics (Agro-Geoinformatics).

Tomer, M.D., et al., 2003. Optimizing the placement of riparian practices in a watershed using terrain analysis. J. Soil Water Conserv. 58 (4), 198–206.

UNDESA, 2012. Sustainable Land Use for the 21st Century (SD21). United Nations Department of Economic and Social Affairs—Division for Sustainable Development.

Valcu-Lisman, A.M., et al., 2016. The optimality of using marginal land for bioenergy crops: tradeoffs between food, fuel, and environmental services. Agric. Resour. Econ. Rev. 45 (2), 217–245.

VanMeerbeek, K., et al., 2015. Biomass of invasive plant species as a potential feedstock for bioenergy production. Biofuels Bioprod. Biorefin. 9, 273–282.

Viator, R.P., et al., 2012. Screening for tolerance to periodic flooding for cane grown for sucrose and bioenergy. Biomass Bioenergy 44, 56–63.

Wicke, B., 2011. Bioenergy Production on Degraded and Marginal Land: Assessing Its Potentials, Economic Performance, and Environmental Impacts for Different Settings and Geographical Scales (Ph.D. Thesis). Faculty of Science, Copernicus Institute, Group Science, Technology and Society, Utrecht Universit.

Williams, C.L., et al., 2013. Grass-shed: place and process for catalyzing perennial grass bioeconomies and their potential multiple benefits. J. Soil Water Conserv. 68 (6), 141A–146A.

Williams, J.D., et al., 2015. Biofuel feedstock production potential in stream buffers of the inland Pacific Northwest: productivity and management issues with invasive plants. J. Soil Water Conserv. 70 (3), 156–169.

Xue, S., et al., 2016. Assesment of the production potentials of Miscanthus on marginal land in China. Renew. Sustain. Energy Rev. 54, 932–943.

Zhao, X., et al., 2014. The viability of biofuel production on urban marginal land: an analysis of metal contaminants and energy balance for Pittsburgh's Sunflower Gardens. Landsc. Urban Plan. 124, 22–33.

Zhuang, D., et al., 2011. Assessment of bioenergy potential on marginal land in China. Renew. Sustain. Energy Rev. 15 (2), 1050–1056.

CHAPTER 7

Implications of biofuel production on direct and indirect land use change: Evidence from Brazil

Amani Elobeid*, Marcelo M.R. Moreira†, Cicero Zanetti de Lima‡, Miguel Carriquiry§, Leila Harfuch†

*Department of Economics and Center for Agricultural and Rural Development, Iowa State University, Ames, IA, United States
†Agroicone, Sao Paulo, Brazil
‡Purdue University, West Lafayette, Indiana, USA
§Department of Economics, Institute of Economics, University of the Republic, Montevideo, Uruguay

Contents

1. Introduction

Land use change, its drivers, and its consequences have been discussed extensively. Drivers such as increases in human population with additional needs for food, changes in the types of food as wealth and urbanization rates increase, demand for energy and fiber, and enhanced transportation and the development of roads (among others) have all been cited as causes of deforestation. Biofuels have gained notoriety in driving land use change and have also been singled out as an important source of deforestation, as grains and oils traditionally used for food are diverted toward bioenergy, spurring the need to expand production to satisfy nutritional requirements. Additional production can be attained by increasing the amount of land used by obtaining higher yields per unit of land or by a combination of the two.

Biofuels, Bioenergy and Food Security
https://doi.org/10.1016/B978-0-12-803954-0.00007-3

Biofuels therefore have direct land use changes as well as indirect land use changes. While direct effects can be captured by knowing how much feedstocks are needed per unit of bioenergy produced and feedstock yields per unit of land, indirect land use changes are much more difficult to assess. They are highly dependent on the interactions of yields of different crops and locations as well as on possibilities for substitution on both the demand and supply sides. Early work recognized the importance of accounting for indirect land use change when assessing the environmental credentials of different types of biofuel pathways (Searchinger et al., 2008). While later softened, this well-cited paper raised the point that accounting for indirect (both at national and global levels) land use change as a result of biofuel expansion was needed because it had important impacts on both the level of greenhouse gas (GHG) emissions and other environmental metrics as well as on commodity prices and food markets. Recognizing its importance, regulators both in the United States and the European Union were tasked to conduct assessments that included direct and indirect land use change to define the types of biofuel pathways that were eligible for support and to be counted toward mandated consumption levels (USDA, 2017; CARB, 2015; Valin et al., 2015).

This chapter is organized as follows. A brief overview of the literature relating to direct and indirect land use change is provided, followed by consideration of land use change for policy implementation. Several factors relevant for the analysis of land use change are then illustrated using recent transitions in Brazil. Final remarks close the chapter.

2. Overview of direct versus indirect land use change

The literature on the impact of biofuels on land use and the environment has evolved over time. Initially, studies accounted for only the direct land use impacts of biofuel production in terms of increased acreage of agricultural feedstock for biofuels relative to the production of fuel from petroleum (Commission of the European Communities, 2006; Farrel et al., 2006; Macedo et al., 2004; Wang et al., 2007; Wang et al., 1999). At that time, the general consensus was that replacing fossil fuels with biofuels would lead to lower GHG emissions. Accounting for only direct land use impacts, producing ethanol from corn led to a saving of GHG emissions when compared to fossil fuels. Hill et al. (2006) estimated a 12% reduction in GHG emissions from the production and combustion of ethanol. Estimates by Wang et al. (2007) averaged higher, at about a 20% saving of GHG emissions relative to fossil fuels. However, studies that took into account indirect land use changes showed increased carbon emissions as land expansion included the clearing of rainforests and grasslands thus releasing carbon.

Subsequent studies, more prominently by Searchinger et al. (2008) and Fargione et al. (2008), found that indirect land use changes from biofuel production resulted in increased carbon emissions. Searchinger's study estimated that the production of corn-based ethanol, when including indirect land use changes, led to a doubling of GHG emissions over 30 years. Ethanol produced from sugarcane also increased emissions but by less than corn. Fargione et al. (2008) found that biofuels produced from food crops resulted in higher GHG emissions, while biofuels produced from biomass or from perennial crops would have no carbon debt and led to GHG savings.

While Searchinger's study highlighted the importance of considering indirect land use changes in the estimation of GHG impacts of biofuel production, the study was criticized for not taking into account the increase in land productivity induced by increased demand and prices, and the use of perennial energy crops on marginal land that did not require the conversion of natural lands to agricultural production (Wang and Haq, 2008; Mathews and Tan, 2009). The early analysis by Searchinger et al. (2008) and Fargione et al. (2008), and later analyses, which took into account land intensification (Hertel et al., 2010; Dumortier et al., 2011), have shown the critical role of land use change in determining lifecycle emission of biofuels.

A review of more recent studies has continued to account for indirect land use changes although there has been a wide range in results based on modeling assumptions (Finkbeiner, 2014; Ahlgren and Di Lucia, 2014). Taheripour and Tyner (2013) found that, in assessing the global impacts of US ethanol expansion, model modifications capturing more recent data resulted in less expansion of cropland, less conversion of forestland, and a decline in land use emissions by 18%.[1] More recent studies have emphasized the importance of land intensification in terms of double cropping and increases in technology-driven and market-driven (price-response) land productivity in assessing land use change (Taheripour et al., 2017a,b; Byerlee et al., 2014). Results showed lower global land use change and emission values from biofuel expansion (Taheripour et al., 2017b).

3. Land use change considerations for biofuel policy implementation

Changes in relative prices between commodities in general and as a result of the expansion of biofuel demand lead to land use change. In particular, higher prices provide incentives for producers to bring more land into agricultural production, and to reallocate land among different agricultural activities. Higher prices are observed not only in countries that increase their feedstock needs because of higher biofuel production and/or consumption but also globally. These global changes in land use may have nontrivial implications for carbon release, posing challenges for policy implementation and fueling the policy debate. In this line, regulations on biofuels require reporting of GHG emission

---

[1] The authors modified the Global Trade Analysis Project model to account for regional responses in terms of the extent and location of land use change as well as updating rates of land conversion to forestland and pasture (Taheripour and Tyner, 2013).

reductions related to feedstock-specific biofuels (Panichelli and Gnansounou, 2015; Warner et al., 2014; Khanna et al., 2017).

Land use changes both direct and indirect are important. Policies such as the USA's Energy Independence and Security Act of 2007 and the European Renewable Fuels Directives indicate that biofuels need to achieve certain targets in terms of lifecycle GHG emission reductions (Hennecke et al., 2013).

There is a large variation in terms of results of lifecycle analysis from different models and methods. Differences at the cultivation and production stages are evident across analyses because a given feedstock can be produced through very distinct practices by producers even within a country (Hennecke et al., 2013). Cross-country differences are even larger in terms of practices, yields achieved on agricultural and newly converted lands, and land uses being displaced (e.g., unused cropland vs. natural forests or peatlands) (Edwards et al., 2010; Babcock, 2015; Plevin et al., 2015).

Some authors conclude that there is a need for harmonization in the calculation of GHG emissions from biofuels (Hennecke et al., 2013; Warner et al., 2014). As indicated, agricultural production, land use needed for a given amount of feedstock, intensity of land use (e.g., possibility for double cropping), and land uses displaced are location specific and can result in very different calculated emission levels. This is important because a regulatory gap is created that needs to be considered for correct implementation of policies, leveling the field, and creating correct incentives to improve the GHG credentials of different biofuels.

4. Evidence from Brazil

To properly assess direct and indirect land use changes resulting from biofuels, it is important to reflect on how agricultural production has responded to price increases from increased demand for biofuel feedstock. This section provides an illustration from Brazil, namely, agricultural production trends, observed land use change based on land intensification (double cropping, increased yields), and extensification (land expansion) in response to price signals.

4.1 Agriculture intensification: The case of double cropping and soybean expansion

Double cropping means planting several crops in the same area and in the same crop year so that the same land is used to generate more than one crop per year. In Brazilian agriculture, double cropping is practiced for maize, peanuts, potatoes, and beans. In some regions, such as Southeast and Northeast Cerrado (MATOPIBA[2]), potatoes and beans

---

[2] Confluence of the States of Maranhão, Tocantins, Piauí, and Bahia. The MATOPIBA is the new agricultural frontier in Brazil.

may even have some areas with a third crop in the same crop year. However, double cropping in maize production represents the most important double cropping in the country. Over the past 10 years, planting summer soybeans (rainy season) with a winter crop of maize has become well established in some regions—South (Paraná State) and Center-West (Mato Grosso)—and is therefore a key driver in the expansion of maize production.

There are two main determinants for such dynamics: the no-till practice for soybean production, which has decreased the time between the harvest of summer soybeans and the planting of maize, and the development of herbicide resistance varieties of maize, high-quality inputs, and technical improvements, which have made it easier to plant the crop directly after soybeans. As presented in Fig. 7.1, the result is an increase in double cropping area in response to higher agricultural prices, thus increasing total production of maize without increasing land use or land use conversion.

Even though the shorter planting period for maize represents more risk compared to traditional maize systems, this process has been increasing significantly across different regions in Brazil. Table 7.1 shows that in the last 20 years, the annual growth rates of planted area of first and second crops of maize were −4.0% and 9.5%, respectively. In the Center-West (Cerrado) and North (Amazon) regions, the annual growth rate reaches 12.8% and 14.9%, respectively. The production growth of the second crop of maize more than compensates for the decline in the first crop as shown in Table 7.1. The second crop of maize is very sensitive to price and profitability, markets, and land use changes.

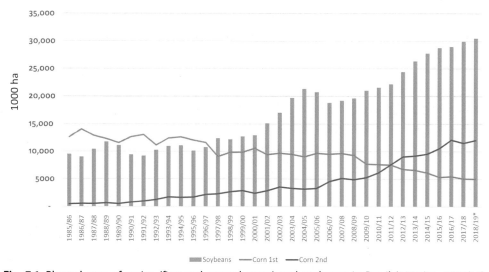

Fig. 7.1 Planted area of maize (first and second crops) and soybeans in Brazil (1985/86–2015/16). *(From Companhia Nacional de Abastecimento (CONAB)—Série Histórica das Safras, 2017. Available at: https://www.conab.gov.br/info-agro/safras/serie-historica-das-safras?start=20).*

Table 7.1 Growth rates of planted area and volume of production of maize (first and second crops) and soybeans (1996–2016)

| | Maize | | | | Soybean | |
| | Planted area | | Volume of production | | | |
	First crop (%)	Second crop (%)	First crop (%)	Second crop (%)	Planted area (%)	Volume of production (%)
Brazil	−4.0	9.5	−0.6	13.1	5.2	6.8
South	−5.2	7.0	−0.3	11.1	4.0	6.0
Southeast	−3.0	4.0	0.5	5.5	3.9	6.2
Center-West	−6.8	12.8	−3.8	16.3	6.0	7.5
North	−4.5	14.9	−2.2	19.5	34.8	35.5
MATOPIBA[a]	−1.2	3.8	3.4	5.5	10.2	10.4
Northeast	−3.9	–	−8.9	–	–	–

[a]The MATOPIBA region is made up of the acronyms of Maranhão (MA), Piauí (PI), Tocantins (TO), and Bahia (BA), which is considered the new agricultural frontier in Brazil.
From Companhia Nacional de Abastecimento (CONAB)—Série Histórica das Safras, 2017. Available at: https://www.conab.gov.br/info-agro/safras/serie-historica-das-safras?start=20.

In terms of soybean planted area and volume of production, the annual growth rates between the 1995/96 and 2015/16 crop years were 5.2% and 6.8%, respectively. These rates are lower than the rates for maize. However, the recent expansion of soybeans, especially in the Cerrado biome, allows for the expansion of the second crop of maize, as shown in Table 7.1. Additionally, soybeans represent 90% (15.6 million hectares [Mha]) of all agriculture in the Cerrado, and in the last crop year more than half (52%) of the soybeans produced in Brazil were concentrated in the Cerrado.

If a particular biofuel crop requires more land, the maize producers would be willing to reduce the area of the first crop of maize and expand the second crop area to adjust land market fluctuations and increase profitability. Increasing double cropping area is one of the farmers' first responses to changes in crop prices. As in Brazil, double cropping in the United States involves soybean production.

A recent comprehensive study on the Cerrado biome and soybean production has shown that there is a potential for Brazilian agriculture to expand soybean production to areas previously occupied by pastures without the need of further deforestation (Filho and Costa, 2016). In the last decade, the agricultural area of Cerrado expanded 87%, and around 70% of the agricultural expansion occurred in pasture or areas with other agricultural crops. The exception is the MATOPIBA region, which has had the greatest expansion over native vegetation areas.

The research also highlights that 33.4 Mha of anthropized areas (areas with high, medium, and low suitability for agriculture without altitude and slope restrictions) would be suitable for conversion into grain production, and 4.2 Mha of this total area are in the MATOPIBA region. Additionally, sustainable cattle ranching intensification initiatives are extremely relevant and strategic to accommodating market changes without losses to the remaining vegetation. This process is already under way in the consolidated agricultural frontier (outside MATOPIBA). As a result, the region has great potential to promote the transition between cattle and soybeans over pasture areas.

4.2 Livestock intensification: Pastureland and integrated systems

Pastureland is around 170 Mha according to the Image Processing and Geoprocessing Lab, Federal University of Goiás (LAPIG–UFG), representing 70% of Brazilian agricultural area. However, the proportions of natural and planted pastureland have changed dramatically over time. Natural pasture was replaced by more profitable planted pasture areas. Planted pasture reached its peak in 1985 (179 Mha), after which pastureland areas declined due to abandonment or shifts to croplands. During the period between 1985 and 2010, planted pasture expanded in the North region following the main rivers and roads (Dias et al., 2016).

In response to market forces, livestock yields increased 129% between 1996 and 2016 (GTPS, 2017). This improvement allowed pastureland areas to decrease by about 20.6 Mha, while, at the same time, meat production increased (Fig. 7.2). Brazilian livestock systems have demonstrated a high endogenous potential to intensify production and absorb any marginal increase in crop area (Nepstad et al., 2014; Latawiec et al., 2014; Strassburg et al., 2014; Dias et al., 2016), which could come from biofuels.

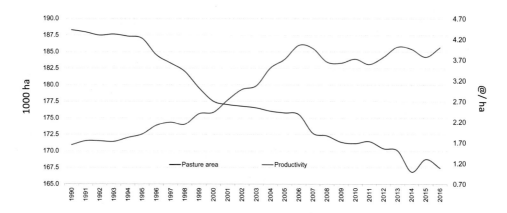

Pasture area and yield in legal Amazon

Fig. 7.2 Pasture area (Mha) and livestock productivity (@/ha) (approximately 15 kg per hectare) in Brazil (1990–2016). *(Adapted from Grupo de Trabalho da Pecuária Sustentável (GTPS). Brazil: Agro-Environmental Potency Challenges of the Planet and Mankind. Available at http://gtps.org.br/biblioteca/). Used with permission from Athenagro Consultoria.*

There are around 30 Mha of pastureland with some level of degradation in Brazil. Although this number is far from being a consensus in the literature due to different definitions of degradation levels and livestock productivity, it is evident that there is a large amount of degraded pastures that could be converted to more efficient use.

Recovered pastureland reduces carbon dioxide (CO_2) emissions by at least 60% in a production system and increases biomass production (Kurihara et al., 1999). Nutrient replacement in the pasture improves animal diet quality. This reduces the time of slaughter and the emissions of methane gas (CH_4) by enteric fermentation (Assad, 2015), which is not currently accounted for in land use change estimation. It also reduces the pressure to convert natural areas into pasture. When compared to degraded pasture, recovered areas provide a higher carbon stock to the system, since there is an accumulation of organic matter in the soil, as well as lower CO_2 losses (32.3 kg CO_2 equivalent per live weight gain for degraded pasture, and 9.8 kg CO_2 equivalent per live weight gain for recovered areas) to the atmosphere. This estimation includes land use change and agricultural emissions factors.

Several studies have shown that policy-driven intensification could reduce pasture areas in Brazil promoting land sparing and reducing the pressure on natural forests and natural areas (Cohn et al., 2014; Strassburg et al., 2014; Silva et al., 2017; Harfuch et al., 2016). The importance of agriculture in mitigating GHG emissions was reinforced in the Brazilian Nationally Determined Contribution (NDC), which additionally aims to recover 15 Mha of degraded pasture, and increase the adoption of integrated systems (ISs) by 5 Mha between 2020 and 2030.

The expansion of soybean production, the degradation of large areas due to livestock ranching, and low livestock productivity in Brazil provide the catalyst for the development of new agricultural practices to intensify agricultural yield, especially in the livestock sector (Salton et al., 2014). Agricultural systems that integrate grain production and livestock ranching could be advantageous to both farmers and the environment. ISs could make it possible to recover pasture productivity and increase crop stability at the same time (Moraes et al., 2014; Sá et al., 2017).

Due to its economic and ecological advantages, ISs have been proposed as a strategy to contain agricultural expansion, the degradation of pastures, and the reduction of deforestation (Lima, 2017). Each type of IS—crop-livestock and/or crop-livestock-forestry—brings its combination of benefits to agriculture. ISs are key among sustainable technologies to overcome the problems arising from decades of using farming practices with high environmental impacts. ISs have the potential to mitigate GHG emissions, reduce erosion and fertility losses, reduce the silting of watercourses, and prevent soil and water pollution, among others. The IS can guarantee the sustainable intensification of agriculture, promoting increased production of foods, fibers, and energy associated with the promotion of ecosystem services (Moraes et al., 2017).

Recent research suggests that there are 11.5 Mha with ISs in Brazil (Rede de Fomento ILPF, 2016). Around 8000 grain (soybean and/or maize) and livestock producers were interviewed. The crop-livestock system with the integration of soybean/maize and livestock is well known by both grain and livestock producers, reaching 99% and 82% of adoption rates in each farmer group, respectively. Other important results are the determinants of IS adoption. In general, farmers are adopting IS because of the yield increases per hectare, higher profitability, recovery of pasture capacity, and reduction of environmental impacts. This information supports the role of livestock intensification progress in Brazil, as well as the mitigation potential of pasture recovery and IS technologies. Consequently, both technologies should be considered for a correct and complete policy assessment in Brazil since livestock intensification directly affects the indirect land use change factor.

4.3 Sugarcane yield and area expansion

Sugarcane is the most promising crop in Brazil to meet additional ethanol demand. It is therefore important to show recent sugarcane yields and regional area expansions. Brazil has already reached a yield of 80 tons per hectare (tons/ha) in the Center-South region—the main Brazilian sugarcane producing area, representing in the 2015/16 crop year about 93% of total production (CONAB, 2017). Several projections for Brazilian agriculture have shown that the yield in 2024/25 would be between 74 and 82 tons/ha (OECD/FAO, 2018; FIESP, 2016). The recent observed yield suggests that Brazil has reached these numbers earlier than projected.

Additional sugarcane production will require area expansion. However, the use of marginal land with lower yields for Brazilian sugarcane seems to underestimate the observed yield trends, especially in the Center-West (Cerrado) region. Logistic cost is the main restriction for sugarcane expansion in the region. Technical assistance and planting techniques adapted to regional characteristics are now broadly available in the "new regions" (Center-West). Table 7.2 shows the recent area expansion of sugarcane in Brazil. There is also evidence that sugarcane expansion creates positive spillovers and increases grain productivity as measured in yield per hectare (Assunção et al., 2016).

Another important point arises from the sugarcane agronomy and management decisions that lead to strong yield/price elasticity. Different from annual crops, sugarcane yield is very sensitive to investments in replanting. After replanting, sugarcane yield reaches the highest yield value and then declines in each successive ratoon. Although replanting is very expensive, sugarcane yield from the first cut is almost double that of the fifth cut, as shown in Table 7.3.

Since this decision combines significant investments and huge improvements in yield, agricultural managers always consider the expected prices for their products (e.g., sugar and ethanol) to determine how much of the area should be renewed each year. If prices

Table 7.2 Sugarcane area expansion in Brazil between 2009/10 and 2016/17 (1000 ha)

Regions	2009/10	2010/11	2011/12	2012/13	2013/14	2014/15	2015/16	2016/17	(2016/17–2009/10)
North	17	20	35	42	46	48	51	52	35
Northeast	1083	1113	1115	1083	1030	979	917	866	−216
Center–West	940	1203	1379	1504	1711	1748	1715	1811	871
Southeast	4833	5137	5221	5243	5436	5593	5455	5700	868
South	537	584	613	612	588	636	517	619	82
Brazil	7410	8056	8363	8485	8811	9004	8655	9049	1640

(From Companhia Nacional de Abastecimento (CONAB)—Série Histórica das Safras, 2017. Available at: https://www.conab.gov.br/info-agro/safras/serie-historica-das-safras?start=20).

Table 7.3 Sugarcane yield per year of cutting (ton/ha)

Crop-year	First cut	Second cut	Third cut	Fourth cut	Fifth cut
2007/08	102	90	79	71	66
2008/09	102	86	76	70	65
2009/10	106	91	78	71	65
2010/11	105	89	77	68	64
2011/12	91	78	69	63	59
2012/13	n.a.	n.a.	n.a.	n.a.	n.a.
2013/14	73	86	77	68	64
2014/15	96	87	76	69	64
Average	96	87	76	69	64

n.a., not available.
(From MAPA, 2012. Ministry of Agriculture, Livestock and Food Supply. Sugarcane Productivity Evolution by Cut. Available at: http://www.agricultura.gov.br/arq_editor/file/Desenvolvimento_Sustentavel/Agroenergia/estatisticas/producao/SETEMBRO_2012/evolucao%20podutividade%20cana.pdf. CONAB, 2014. National Company of Food Supply. Conab—Levantamento: Agosto/2014—2° Levantamento da Safra 2014/15. Available at: http://www.agricultura.gov.br/arq_editor/file/camaras_setoriais/Acucar_e_alcool/27RO/App_Safra_27RO_Alcool.pdf).

are high, producers are willing to invest and expect a higher return. This relationship is clear and statistically significant, as shown in Fig. 7.3.

If the price is high, the producer tends to renew a greater share of the sugarcane area earlier, leading to a higher agricultural yield. Therefore the yield increase should be considered as an important way to achieve the additional sugarcane production required in both sugarcane pathways. Similarly, for crop yields, improvements in industrial performance are another aspect that must be accounted for in indirect land use change assessments.

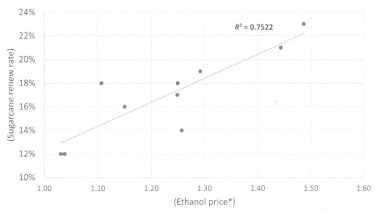

Fig. 7.3 Correlation between ethanol price and sugarcane renew rate. *Real/L (2014 real values). (Based on the sources UNICA (2016). The Brazilian Sugarcane Industry Association. Coletiva de imprensa— Estimativa Safra 2016/2017. Available at: http://unica.com.br/download.php?idSecao=17&id=10968146. CEPEA, 2016. Center for Advanced Studies on Applied Economics. Available at: http://cepea.esalq.usp.br/ etanol/#), accessed in 2016.*

4.4 Land use polices

In the last two decades, Brazil has established and reinforced a set of land use public and private policies, as well as control mechanisms. Nepstad et al. (2014) showed how enforcement of laws and interventions in soybean and beef supply chains, restrictions on access to credit, and expansion of protected areas appear to have contributed to the large decline in Amazon deforestation rates. Among the more significant policies and commitments established in Brazil are Sugarcane and Palm Oil Zoning (ZAE-Cana and ZAE-Palma, respectively), the Low Carbon Agricultural Plan (ABC Plan, in its Portuguese acronym), the voluntary commitment on reducing deforestation, the revision of the Forest Code (including Cadastro Ambiental Rural, CAR, in Portuguese), the Soybean Moratorium, and the commitment on eliminating illegal deforestation.

As part of the National Plan on Climate Change, the Plan for the Protection and Control of Deforestation in the Amazon (PPCDAm) and the Plan for the Protection and Control of Deforestation in the Cerrado (PPCerrado) commit to significant reductions in deforestation rates in the legal Amazon and Cerrado biome (80% and 40% reduction, respectively). Additionally, the government has put in place a set of strategies, such as the creation of new protection areas (with different restriction levels), better interactions between governmental bodies, as well as satellite vigilance to monitor and target specific deforestation focus, with effective results (Nepstad et al., 2014; Assunção et al., 2013). These measures have contributed to lower deforestation rates (Fig. 7.4).

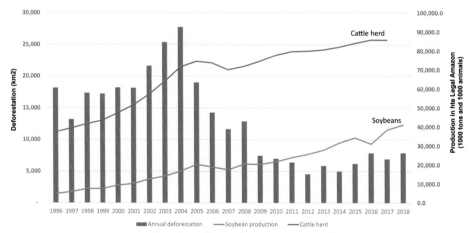

Fig. 7.4 Annual deforestation rate (1000 km²), cattle herd (million animals), and soybean production (million metric tons) in the Legal Amazon between 1996 and 2016. *(From INPE, 2017. PRODES—Monitoramento da Floresta Amazônica Brasileira por Satélite. http://www.obt.inpe.br/OBT/assuntos/programas/amazonia/prodes. IBGE, 2017. Pesquisa Pecuária Municipal. Available at: https://www.ibge.gov.br).*

Launched in 2009, ZAE-Cana restricts the expansion of sugarcane over several land use classes, including areas covered by natural vegetation, the Upper Paraguay Basin Amazon and Pantanal Biomes, and indigenous areas, among others. According to the new criteria, 92.5% of the national territory is considered not suitable for sugarcane plantation having several restrictions for its expansion (EMBRAPA, 2009).

Launched in 2010, Palm Oil Zoning restricts palm oil expansion into native vegetation (EMBRAPA, 2010). To access official credit, oil palm producers should only use land deforested before 2008. Investors are required to adhere to the ZAE-Palma criteria and deforestation laws, which are being strictly monitored and enforced (Brandão and Schoneveld, 2015).

Launched in 2006 and with the participation of major trading companies, the Ministry of Environment, Brazilian Banks, and NGOs, the Soybean Moratorium restricted the purchase of soybeans from deforested areas of the Amazon biome after 2006. The cutoff date was postponed to 2008 to comply with the National Forestry Code. The moratorium has been recognized as an efficient tool to avoid deforestation, which should be maintained (Nepstad et al., 2014; Gibbs et al., 2015). In 2004, up to 30% of the planted soybeans came from recent deforestation in the Amazon. Today, this value is no higher than 1.25% (Greenpeace, 2016).

The Native Vegetation Protection Law No. 12.651/2012 (also known as the Forest Code) governs the use and protection of public and private lands in Brazil. It is a key policy instrument to promote restoration of natural vegetation, curb illegal deforestation, and regulate permitted conversion or legal deforestation. Key elements include the establishment of an Environmental Rural Registry (CAR) and Environmental Compliance Programmes (Programas de Regularização Ambiental, PRAs, in Portuguese), and obligations to keep and restore Permanent Preservation Areas (APPs) and Legal Reserves (LRs). According to Chiavari and Lopes (2017) it is one of the most important pieces of legislation with the potential to drive efficient land use in the country and become an effective tool against climate change.

Fig. 7.5 presents the best representation of land use in Brazil (left) and legal protection of natural vegetation (right). Even though native vegetation forests still account for 67% of the Brazilian territory (followed by pastures, 20%), only a share is not legally protected (28%). Considering the total remaining natural vegetation, around 28% is protected under conservation units or indigenous lands, and 34% is protected within private farmland.

Regarding private land, the Forest Code protects riparian areas (buffers vary from 5 to 500 m) and a percentage of private land native vegetation ranging from 20% to 80% must mandatorily be set aside. As shown in Fig. 7.6 the set-aside areas have a higher C stock, such as the Amazon biome.

It is important to mention that the current Forest Code was revised in 2012 seeking for effective implementation. A programmatic increase in formality of farms has been

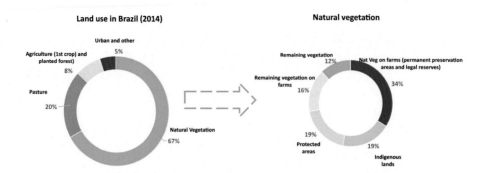

Fig. 7.5 Land use in Brazil and native vegetation. *(From Agroicone based on IBGE-PAM, 2014. Sparovek et al., 2015. A adicionalidade do mecanismo de compensação de reserva legal da lei no 12.651/2012: uma análise da oferta e demanda de cotas de reserva ambiental (Chap. 5). Availabel at: http://www. ipea.gov.br/agencia/images/stories/PDFs/livros/livros/160812_livro_mudancas_codigo_florestal_ brasileiro_cap5.pdf).*

Fig. 7.6 Share of private properties that need to be set aside as Legal Reserves. *Note*: Calculations for all categories considered the best available data in 2014. *LR*, Legal Reserve. *(Used with permission from Agroicone.).*

established with special rules (compensation) for areas deforested before 2008. For the Atlantic Forest biome (tropical forest), restrictions are more stringent than the 20%, once typical forestry python-fisionomes (dense forests) are fully protected.

The Environmental Rural Registry (CAR) is one of the most important features of the new Forest Code. CAR is an electronic (geographical information system-based) national public registry mandatory for all rural properties with the purpose of integrating environmental information, and is composed of a database for control, monitoring, environmental and economic planning, and can also combat deforestation. Registration in the CAR is a mandatory condition for the exercise of several rights, such as obtaining authorization for the suppression of native vegetation and the maintenance of activities in consolidated areas, among others. In addition, all financial institutions will only grant agricultural credit to rural properties registered in the CAR (Silva et al., 2016). Until August 31, 2017, more than 4.3 million rural properties have been registered with a total area of 413 Mha inserted in the database (SFB, 2017), which represents more than 100% of the agricultural area.

Looking forward, the Brazilian NDC is heavily based on reinforcing land use policies jointly with biofuels expansion. Forestry Code implementation and the elimination of illegal deforestation are clearly established. In addition, the current legislation will reinforce its effectiveness.

Given the voluntary commitment made at COP-15,[3] in 2010 Brazil released the Low Carbon Agriculture Plan (ABC Plan, in Portuguese). The ABC Plan is a sectoral plan to mitigate GHG emissions in agriculture, improve efficiency in the use of natural resources, and increase the resilience of productive systems and rural communities, as well as enable the sector to adapt to climate change. The ABC Plan has several actions to be implemented by 2020, such as (1) recover 15 Mha of degraded pasture; (2) increase the adoption of ISs by 4 Mha; (3) increase the no-till system by 8 Mha; (4) increase biological nitrogen fixation by 5.5 Mha, substituting nitrogen fertilizers; (5) increase planted forests by 3 Mha; (6) expand the treatment of animal waste by 4.4 million cubic meters; and (7) encourage the implementation of adaptation actions to climate change, especially to those with GHG mitigation potential.

5. Final remarks

The expansion of biofuel production and consumption as a result of policies and market forces has led to the expansion of land used for agriculture as well as reallocation of land within crops and agricultural activities. This chapter provided a general overview of the implications of the expansion of biofuels and feedstock needs as a source of both direct and indirect land use changes. These changes can occur completely in the country in

[3] Brazil has committed to reduce GHG emissions by 37% below 2005 levels in 2025 and 43% in 2030.

which the biofuel demand arises or could be exported either partially or in full to other countries around the world. The extent of the expansion and reallocation of land around the world, including the geographical zones affected, will depend on the biofuel pathway considered. Other sources of demand for biofuels feedstocks, as well as the possibilities of substitution in both the supply and demand sides, also affect the extent and location of land use change.

Direct and indirect land use changes affect the environmental credentials of biofuels in terms of possible contamination of water and soil resources, and GHG emissions among other impacts. Market-mediated and price implications were also intensely analyzed as part of the food versus fuels debate. In short, it was argued that diverting commodities from food to fuel uses raises prices of grains, oils, and animal-based products, thus increasing the number of people that are food insecure.

The impacts just mentioned were recognized early and policies in several countries required that biofuels meet certain environmental criteria, most prevalently in terms of lifecycle reductions of GHG emissions relative to the fossil fuels that were displaced. Land use changes both direct and indirect were shown to be critical in the assessment of lifecycle GHG emissions of biofuels. This is also an area in which considerable scientific uncertainty remains with different modeling groups and approaches leading to widely diverging assessments as indicated here and highlighted in other chapters of the book (see, e.g., Dumortier, Elobeid, and Carriquiry).

A major difficulty in assessing land use change is the diversity of possibilities in which agricultural activities can be conducted, including production practices, climatic conditions, resource bases, etc. These are all highly specific, location dependent, and reflected and modulated by institutional and cultural aspects. The dynamics of agricultural expansion can also change in short periods of time as they respond to policy, technologies, and market forces. The recent evolution of agricultural activities in Brazil, including zoning restrictions, possibilities for double cropping, and stricter enforcement of policies, all illustrate the difficulties and the need to frequently evaluate and update land use change assessments.

References

Ahlgren, S., Di Lucia, L., 2014. Indirect land use changes of biofuel production—a review of modelling efforts and policy developments in the European Union. Biotechnol. Biofuels 2014 (7), 35. https://doi.org/10.1186/1754-6834-7-35.

Assad, E.D., 2015. Agricultura de Baixa Emissão de Carbono: A Evolução de Um Novo Paradigma Eduardo de Morais Pavão. Observatório ABC. http://bibliotecadigital.fgv.br/dspace/bitstream/handle/10438/15353/AgriculturadebaixaemissãodecarbonoAevoluçãodeumnovoparadigma.pdf?sequence=1&isAllowed=y.

Assunção, J., Gandour, C., Rocha, R., 2013. DETERring deforestation in the Brazilian Amazon: environmental monitoring and law enforcement. Clim. Pol. Initiat.

Assunção, J., Pietracci, B., Souza, P., 2016. Fueling development: sugarcane expansion impacts in Brazil. INPUT-Iniciativa para o Uso da Terra 55.

Babcock, B.A., 2015. Extensive and intensive agricultural supply response. Annu. Rev. Resour. Econ. 7, 333–348.

Brandão, F., Schoneveld, G., 2015. The state of oil palm development in the Brazilian Amazon: trends, value chain dynamics, and business models. In: CIFOR Working Paper. Center for International Forestry Research (CIFOR), Bogor, Indonesia. https://doi.org/10.17528/cifor/005861.

Byerlee, D., Stevenson, J., Villoria, N., 2014. Does intensification slow crop land expansion or encourage deforestation? Glob. Food Sec. 3 (2), 92–98. https://doi.org/10.1016/j.gfs.2014.04.001.

CARB, 2015. Calculating carbon intensity values from indirect land use change of crop-based biofuels. Staff report, California Environmental Protection Agency Air Resources Board.

Chiavari, J., Lopes, C.L., 2017. Forest and Land Use Policies on Private Lands: An International Comparison: Argentina, Brazil, Canada, China, France, Germany, and the United States. Climate Policy Initiative Full report. Available at: https://climatepolicyinitiative.org/wp-content/uploads/2017/10/Full_Report_Forest_and_Land_Use_Policies_on_Private_Lands_-_an_International_Comparison-1.pdf.

Cohn, A.S., Mosnier, A., Havlík, P., Valin, H., Herrero, M., Schmid, E., O'Hare, M., Obersteiner, M., 2014. Cattle ranching intensification in Brazil can reduce global greenhouse gas emissions by sparing land from deforestation. Proc. Natl. Acad. Sci. U. S. A. 111 (20), 7236–7241. https://doi.org/10.1073/pnas.1307163111.

Commission of the European Communities, 2006. Biofuels progress report: Report on the progress made in the use of biofuels and other renewable fuels in the member states of the European Union. (COM(2006) 845 final, Brussels).

Companhia Nacional de Abastecimento (CONAB)—Série Histórica das Safras, 2017. Available at: https://www.conab.gov.br/info-agro/safras/serie-historica-das-safras?start=20.

Dias, L.C.P., Pimenta, F.M., Santos, A.B., Costa, M.H., Ladle, R.J., 2016. Patterns of land use, extensification, and intensification of Brazilian agriculture. Glob. Chang. Biol. 22 (8), 2887–2903. https://doi.org/10.1111/gcb.13314.

Dumortier, J., Hayes, D.J., Carriquiry, M., Dong, F., Du, X., Elobeid, A., Fabiosa, J.F., Tokgöz, S., 2011. Sensitivity of carbon emission estimates from indirect land-use change. Appl. Econ. Perspect. Policy, 1–21. https://doi.org/10.1093/aepp/ppr015.

Edwards, R., Mulligan, D., Marelli, L., 2010. Indirect Land Use Change from Increased Biofuels Demand—Comparison of Models and Results for Marginal Biofuels Production from Different Feedstocks. Publications Office of theEuropean Union. Available at: http://publications.jrc.ec.europa.eu/repository/handle/JRC59771, doi:10.2788/54137.

EMBRAPA, 2009. Zoneamento Agroecológico da Cana-de-Açúcar. Available at: https://www.embrapa.br/busca-de-solucoes-tecnologicas/-/produto-servico/1249/zoneamento-agroecologico-da-cana-de-acucar.

EMBRAPA, 2010. Zoneamento Agroecológico do dendezeiro para as áreas desmatadas da Amazônia Legal. Available at: https://www.embrapa.br/en/busca-de-solucoes-tecnologicas/-/produto-servico/1248/zoneamento-agroecologico-do-dendezeiro-para-as-areas-desmatadas-da-amazonia-legal.

Fargione, J., Hill, J., Tilman, D., Polasky, S., Hawthorne, P., 2008. Land clearing and the biofuel carbon debt. Science 39 (5867), 1235–1238. https://doi.org/10.1126/science.1152747.

Farrel, A.E., Plevin, R.J., Turner, B.T., Jones, A.D., O'Hare, M., Kammen, D.M., 2006. Ethanol can contribute to energy and environmental goals. Science 311 (5760), 506–508. https://doi.org/10.1126/science.1121416.

FIESP, 2016. Departamento do Agronegócio, Outlook Fiesp, 2025. Projeções Para o Agronegócio Brasileiro/FIESP. FIESP, São Paulo.

Filho, A.C., Costa, K., 2016. The Expansion of Soybean Production in the Cerrado. São Paulo, http://www.inputbrasil.org.

Finkbeiner, M., 2014. Indirect land use change—help beyond the hype? Biomass Bioenergy 62, 218–221.

Gibbs, H.K., Rausch, L., Munger, J., Schelly, I., Morton, D.C., Noojipady, P., Soares-Filho, B., Barreto, P., Micol, L., Walker, N.F., 2015. Brazil's Soy Moratorium. Science 347 (6220). 377 LP-378, http://science.sciencemag.org/content/347/6220/377.abstract.

Greenpeace, 2016. The Soy Moratorium, 10 Years on: How One Commitment Is Stopping Amazon Destruction. http://www.greenpeace.org/international/en/news/Blogs/makingwaves/the-soy-moratorium-10-year-anniversary-stopping-amazon-destruction/blog/57127/.

GTPS, 2017. Brazilian Roundtable on Sustainable Livestock. Brazilian Livestock and its contribution to Sustainable Development. http://agroicone.com.br/uploads/2015/12/position-paper.pdf.

Harfuch, L., Lima, R., Bachion, L.C., Moreira, M.M.R., Antoniazzi, L., Palauro, G., Kimura, W., et al., 2016. Cattle Ranching Intensification as a Key Role on Sustainable Agriculture Expansion in Brazil. São Paulo, http://www.inputbrasil.org.

Hennecke, A.M., Faist, M., Reinhardt, J., Junquera, V., Neeft, J., Fehrenbach, H., 2013. Biofuel greenhouse gas calculations under the European Renewable Energy Directive—a comparison of the BioGrace tool vs. the tool of the Roundtable on Sustainable Biofuels. Appl. Energy 102, 55–62.

Hertel, T.W., Golub, A.A., Jones, A.D., O'Hare, M., Plevin, R.J., Kammen, D.M., 2010. Effects of US maize ethanol on global land use and greenhouse gas emissions: estimating market-mediated responses. Bioscience 60 (3), 223–231. https://doi.org/10.1525/bio.2010.60.3.8.

Hill, J., Nelson, E., Tilman, D., Polasky, S., Tiffany, D., 2006. Environmental, economic, and energetic costs and benefits of biodiesel and ethanol biofuels. Proc. Natl. Acad. Sci. U. S. A. 103 (30), 11206–11210.

Khanna, M., Wang, W., Hudiburg, T.W., DeLucia, E.H., 2017. The social inefficiency of regulating indirect land use change due to biofuels. Nat. Commun. 8.

Kurihara, M., Magner, T., Hunter, R.A., Mccrabb, G.J., 1999. Methane production and energy partition of cattle in the tropics. Br. J. Nutr. 81 (1), 227–234. https://doi.org/10.1017/S0007114599000422.

Latawiec, A.E., Strassburg, B.N.S., Valentim, J.F., Ramos, F., 2014. Intensification of cattle ranching production systems: socioeconomic and environmental synergies and risks in Brazil. Animal 8 (8), 1255–1263.

Lima, C.Z., 2017. Impacts of Low-Carbon Agriculture in Brazil: A CGE Application. Federal University of Vicosa.

Macedo, I., Leal, M.L.V., da Silva, J.A.R., 2004. Assessment of Greenhouse Gas Emissions in the Production and Use of Fuel Ethanol in Brazil. Secretariat of the Environment, Government of the State of São Paulo.

Mathews, J.A., Tan, H., 2009. Biofuels and indirect land use change effects: the debate continues. Biofuels Bioprod. Biorefin. 3 (3), 305–317. https://doi.org/10.1002/bbb.147.

Moraes, A.D., Carvalho, P.C.D.F., Brasil, S., Lustosa, C., Lang, C.R., Deiss, L., 2014. Research on integrated crop-livestock systems in Brazil. Rev. Ciênc. Agron. 45 (5), 1024–1031. www.ccarevista.ufc.br.

Moraes, A., de Faccio Carvalho, P.C., Pelissari, A., Alves, S.J., Lang, C.R., 2017. Sistemas de Integração Lavoura-Pecuária No Subtrópico Da América Do Sul: Exemplos Do Sul Do Brasil. https://www.researchgate.net/profile/Paulo_De_Faccio_Carvalho2/publication/264846535_Sistemas_de_integracao_lavoura-pecuaria_no_Subtropico_da_America_do_Sul_Exemplos_do_Sul_do_Brasil/links/53fdff4d0cf2364ccc0a0272/Sistemas-de-integracao-lavoura-pecuaria-n. (Accessed 4 May 2019).

Nepstad, D., McGrath, D., Stickler, C., Alencar, A., Azevedo, A., Swette, B., Bezerra, T., et al., 2014. Slowing Amazon deforestation through public policy and interventions in beef and soy supply chains. Science 344 (6188), 1118–1123. http://science.sciencemag.org/content/344/6188/1118.abstract.

OECD/FAO, 2018. OECD-FAO Agricultural Outlook 2018–2027, OECD Publishing, Paris/Food and Agriculture Organization of the United Nations, Rome. Available at: https://doi.org/10.1787/agr_outlook-2018-en.

Panichelli, L., Gnansounou, E., 2015. Impact of agricultural-based biofuel production on greenhouse gas emissions from land-use change: key modelling choices. Renew. Sust. Energ. Rev. 42, 344–360.

Plevin, R.J., Beckman, J., Golub, A.A., Witcover, J., O'Hare, M., 2015. Carbon accounting and economic model uncertainty of emissions from biofuels-induced land use change. Environ. Sci. Technol. 49 (5), 2656–2664.

Rede de Fomento ILPF, 2016. Avaliação Da Adoção de Sistemas de Integração Lavoura-Pecuária-Floresta (ILPF) No Brasil.

Sá, J.C.M., Lal, R., Cerri, C.C., Lorenz, K., Hungria, M., de Faccio Carvalho, P.C., 2017. Low-carbon agriculture in South America to mitigate global climate change and advance food security. Environ. Int. 98, 102–112. https://doi.org/10.1016/j.envint.2016.10.020. Elsevier Ltd.

Salton, J.C., Mercante, F.M., Tomazi, M., Zanatta, J.A., Concenço, G., Silva, W.M., Retore, M., 2014. Integrated crop-livestock system in tropical Brazil: toward a sustainable production system. Agric. Ecosyst. Environ. 190, 70–79. https://doi.org/10.1016/j.agee.2013.09.023.

Searchinger, T., Heimlich, R., Houghton, R.A., Dong, F., Elobeid, A., Fabiosa, J., et al., 2008. Use of U.S. croplands for biofuels increases greenhouse gases through emissions from land-use change. Science 319 (5867), 1238–1240. https://doi.org/10.1126/science.1151861.

SFB, 2017. Serviço Florestal Brasileiro - Números Do Cadastro Ambiental Rural. http://www.florestal.gov.br/modulo-de-relatorios.

Silva, A.P.M., Marques, H.R., Sambuichi, R.H.R., 2016. Mudanças No Código Florestal Brasileiro: Desafios Para Implementação Da Nova Lei. Rio de Janeiro.

Silva, J.G., Ruviaro, C.F., de Souza Ferreira Filho, J.B., 2017. Livestock intensification as a climate policy: lessons from the Brazilian case. Land Use Policy 62, 232–245. https://doi.org/10.1016/j.landusepol.2016.12.025.

Strassburg, B.B.N., Latawiec, A.E., Barioni, L.G., Nobre, C.A., da Silva, V.P., Valentim, J.F., Vianna, M., Assad, E.D., 2014. When enough should be enough: improving the use of current agricultural lands could meet production demands and spare natural habitats in Brazil. Glob. Environ. Chang. 28 (1), 84–97. https://doi.org/10.1016/j.gloenvcha.2014.06.001. Elsevier Ltd.

Taheripour, F., Tyner, W.E., 2013. Biofuels and land use change: applying recent evidence to model estimates. Appl. Sci. 3, 14–38. https://doi.org/10.3390/app3010014.

Taheripour, F., Cui, H., Tyner, W.E., 2017a. An exploration of agricultural land use change at intensive and extensive margins: implications for biofuel-induced land use change modeling. In: Qin, Z., Mishra, U., Hastings, A. (Eds.), Bioenergy and Land Use Change. American Geophysical Union, pp. 19–37. https://doi.org/10.1002/9781119297376.ch2. Chapter 2.

Taheripour, F., Zhao, X., Tyner, W.E., 2017b. The impact of considering land intensification and updated data on biofuels land use change and emissions estimates. Biotechnol. Biofuels. 10. https://doi.org/10.1186/s13068-017-0877-y.

USDA, 2017. A life-cycle analysis of the greenhouse gas emissions of corn-based ethanol. Report prepared by icf under USDA contract no. ag-3142-d-16-0243, U.S. Department of Agriculture.

Valin, H., Peters, D., van den Berg, M., Frank, S., Havlik, P., Forsell, N., Hamelinck, C., Pirker, J., et al., 2015. The Land Use Change Impact of Biofuels Consumed in the EU: Quantification of Area and Greenhouse Gas Impacts. ECOFYS Netherlands B.V., Utrecht, Netherlands. doi:BIENL13120.

Wang, M., Haq, Z., 2008. Letter to science: response to the article by Searchinger et al. in the February 7, 2008, Sciencexpress, Use of U.S. Croplands for Biofuels Increases Greenhouse Gases through Emissions from Land Use Change. https://www.afdc.energy.gov/pdfs/letter_to_science_anldoe_03_14_08.pdf. (Accessed May 2018).

Wang, M., Saricks, C., Santini, D., 1999. Effects of fuel ethanol use on fuel-cycle energy and greenhouse gas emissions. Center for Transportation Research, Energy Systems Division, Argonne National Laboratory, Argonne, IL.

Wang, M., Wu, M., Huo, H., 2007. Life-cycle energy and greenhouse gas emission impacts of different corn ethanol plant types. Environ. Res. Lett. 2.

Warner, E., Zhang, Y., Inman, D., Heath, G., 2014. Challenges in the estimation of greenhouse gas emissions from biofuel-induced global land-use change. Biofuels, Bioproducts and Biorefining 8 (1), 114–125. https://doi.org/10.1002/bbb.1434.

Further reading

Herrero, M., Henderson, B., Havlík, P., Thornton, P.K., Conant, R.T., Smith, P., Wirsenius, S., et al., 2016. Greenhouse gas mitigation potentials in the livestock sector. Nat. Clim. Chang. 6 (5), 452–461. https://doi.org/10.1038/nclimate2925. Nature Publishing Group.

Plevin, R.J., Gibbs, H.K., Duffy, J., Yui, S., Yeh, S., 2014. Agro-Ecological Zone Emission Factor (AEZ-EF) Model (v47). GTAP (Global Trade Analysis Project), pp. 1–48.

Stevanović, M., Popp, A., Bodirsky, B.L., Humpenöder, F., Müller, C., Weindl, I., Dietrich, J.P., et al., 2017. Mitigation strategies for greenhouse gas emissions from agriculture and land-use change: consequences for food prices. Environ. Sci. Technol. 51 (1), 365–374. https://doi.org/10.1021/acs.est.6b04291.

CHAPTER 8

Biofuels' contribution to date to greenhouse gas emission savings

Claire Palandri*, Céline Giner[†], Deepayan Debnath[‡]
*School of International and Public Affairs, Columbia University, New York, NY, United States
[†]Trade and Agriculture Directorate, Organisation for Economic Co-operation and Development, Paris, France
[‡]Food and Agricultural Policy Research Institute, University of Missouri, Columbia, MO, United States

Contents

Acronyms

2DS	Two degrees scenario
BAU	Business-as-usual
CO2e	Carbon dioxide-equivalent
ETP	Energy Technology Perspectives report
FAO	Food and Agriculture Organization of the United Nations
GHG	Greenhouse gas
IEA	International Energy Agency
ILUC	Indirect land use change
IPCC	Intergovernmental Panel on Climate Change
LCA	Lifecycle assessment
LUC	Land use change
mln	Million
OECD	Organization for Economic Cooperation and Development
RFTO	Renewable transport fuel obligation
UNFCCC	United Nations Framework Convention on Climate Change
WTW	Well-to-wheels

Biofuels, Bioenergy and Food Security
https://doi.org/10.1016/B978-0-12-803954-0.00008-5

1. Introduction

By providing substitutes for fossil fuels, biofuels have the potential to contribute to the mitigation of greenhouse gas (GHG) emissions in the transport sector. The magnitude of this contribution depends on the feedstock of origin, the conversion process, and the fuel they ultimately replace.

In the context of increasing global mitigation goals, ambitious policies directly addressing climate and energy objectives are being developed and implemented across the world. As the transport sector represents a significant part of global emissions—accounting for 23% of global energy-related CO_2 emissions in 2015 (IEA, 2017a,b)—and transport demand is expected to increase in the near future, developing strategies to work toward the decarbonization of the transport sector is of the utmost importance. The potential of biofuels to mitigate emissions is therefore of great interest.

However, the actual contribution of biofuels to the reduction of GHG emissions in the transport sector is not as evident as it may have appeared to be in the early 2000s. Over the last 10 years, research on the emissions associated with direct and indirect land use changes (LUCs) caused by the production of biofuels has generated a vigorous debate about the extent to which biofuels actually lead to emission reductions (Khanna and Crago, 2012). This chapter addresses the following questions: Which categories of biofuel have the potential to lead to significant GHG emission savings? What amount of transport-related emissions has been avoided by the use of biofuels up to 2017 (noting that consumption has remained dominated by conventional biofuels)? What is their expected evolution in a business-as-usual (BAU) scenario?

This chapter is organized as follows. The first section contains a detailed explanation of how emission factors for the main categories of biofuels can be estimated and provides such estimates, informed by a review of the existing literature. These estimates are then matched against those of the corresponding fossil fuels, laying the ground for assessing the actual GHG savings achieved by the consumption of biofuels. The second section pursues this by adapting a partial equilibrium model of world agricultural markets. Results from the modeling exercise provide a picture of total historical emissions savings until 2017 and medium-term projections in a BAU scenario.

2. Estimating the carbon intensity of biofuels

Biofuels were originally considered by many countries as carbon neutral. The argument was that although the combustion of their hydrocarbon chains leads to the emission of GHGs, the feedstocks from which biofuels are produced have previously incorporated carbon from the atmosphere during their growth. The amount of carbon released is therefore equal to the amount initially absorbed, resulting in neutral net carbon intensity.

This argument does not hold when one expands the scope of study, such as when conducting a lifecycle assessment (LCA)[1] of biofuels. LCAs consist of an assessment of the environmental impacts of a product or a service that are introduced at each step along the value chain. This approach takes into account all the impacts accumulated at each step of the life of the product, from the production of the raw material to the disposal of the final product, by conducting a thorough analysis of each input and output. LCAs consider a broad range of environmental criteria, such as water pollution, emission of fine particles, biodiversity loss or gain, etc. Analyzing the carbon footprint of a product corresponds essentially to conducting an LCA restricted to one category of environmental impacts, namely GHG emissions.

In the case of biofuels, important sources of emissions are found at every step of the product's value chain, notably during plant growth (which often entails the use of fertilizers), crop harvest (use of fuel by agricultural harvesting machinery and vehicles), transportation of the feedstock to the processing plant (fuel consumption), conversion process (energy use and addition of inputs), distribution to the fuel terminal (consumption of fuel during transport), and finally combustion in the vehicle's engine.[2,3] The aggregation of these different sources leads to positive amounts of GHG emissions. Furthermore, as will be discussed in the next section, expanding the scope further by including the impacts of LUCs induced by the cultivation of crops for the production of biofuels greatly exacerbates these results.

However, despite their positive carbon footprint, biofuels may provide "GHG emission savings" in comparison with the default alternative, namely fossil fuels. In that context, it is important to assess the actual contribution of biofuels to the reduction of GHG emissions and conduct this exercise for the various biofuel pathways.

2.1 Insights from LCA analyses and LUC modeling

A vast range of studies have undertaken estimations of the GHG emissions associated with the use of biofuels. A review of the existing literature provides valuable insights on the necessity to differentiate the multiple biofuel conversion pathways.[4] However, these studies rely on various assumptions, scopes, and estimation methods. To obtain estimates

---

[1] The framework for conducting an LCA has been standardized under ISO 14040. See: https://www.iso.org/standard/37456.html.

[2] Another notion closely related to LCAs is the one of "embodied emissions." The Intergovernmental Panel on Climate Change (IPCC) defines them as "emissions that arise from the production and delivery of a good or service or the build-up of infrastructure. Depending on the chosen system boundaries, upstream emissions are often included (e.g., emissions resulting from the extraction of raw materials)" (Allwood et al., 2014).

[3] The examples provided in parentheses are not exhaustive.

[4] A more extensive discussion of the concept of biofuel pathways is conducted in Chapter 2.

of the emissions associated with the different biofuel feedstocks and conversion processes that are consistent with one another, the assumptions and steps conducted to obtain them must be comparable. To satisfy this, the results presented in the following pages were mostly extracted from comprehensive studies that estimated GHG emissions for a range of biofuel categories.

To assess the GHG emission savings potential of biofuels, we consider separately two components of their carbon intensity: GHGs emitted along the value chain, estimated through an LCA approach, and GHG emissions arising from LUC.

GHG emissions from land use change (LUC)

Emissions from LUC originate from changes in the storage capacity of the two "carbon sinks" that are the soil and the above-ground biomass. Modification of land cover induces changes in carbon storage capacity, which implies net losses or gains of carbon content in these two reservoirs. For example, when a parcel is deforested to allow for the cultivation of crops, the change from high-vegetation to low-vegetation biomass usually results in net carbon emissions to the atmosphere, as crops store less carbon than trees and losses of soil carbon content also ensue.

In the case of biofuels, LUC-related emissions are further divided into two categories. "Direct" emissions result from cropland conversions to crops destined for the production of biofuel feedstocks; "indirect" emissions occur when this land conversion of existing crops leads to the production of food and feed crops on new agricultural land to maintain the level of agricultural production for food and feed production. In the second case, depending on the nature of the previous land cover, the expansion of agricultural land can induce reductions of the carbon storage capacity of biomass and soil, leading to the release of CO_2 into the atmosphere.

The controversy over the impact of biofuel production on LUCs arose in the late 2000s, alongside other concerns regarding the sustainability and potential negative impacts of biofuels.[5] Multiple studies have been launched since 2009 to examine the extent and consequences of indirect LUC, notably in Europe and the United States. The majority of them rely on economic models, either partial equilibrium models specialized in the agricultural sector or general equilibrium models. LUC impacts of biofuels can be estimated by comparing land use types in the baseline situation with a scenario that simulates an increase in biofuel demand. However, this method does not allow for the identification of the part of these LUCs that is caused directly versus indirectly. Many studies based on economic models thus provide total (direct + indirect) LUC estimates.

---

[5] A larger discussion of these developments is provided in Chapter 2.

Estimating emission factors

The carbon intensity of biofuels can be represented by an "emission factor,"[6] expressed in $kgCO_2e$ per unit of volume, mass, or energy content. Let us note that for the purpose of this chapter, we will not consider specific GHGs individually but rather the aggregate of all GHGs identified, expressed in CO_2-equivalent[7] (CO_2e).

In this section, the emissions resulting from biofuel consumption are calculated for ethanol and biodiesel separately and for the major biofuel producers and consumers. To do so, emission factors are assessed independently for the various biofuel pathways. Results from a literature review enabled the attribution of two complementary emission factors for each feedstock category:

- an emission factor corresponding to the LCA of the product (accounting for the cultivation, processing, transport, and distribution of biofuels, and excluding LUC considerations);
- an emission factor accounting for the LUC effects.

The majority of the scientific community seems to agree about the existence of the LUC effect and the significance of resulting emissions. A study conducted in 2012 by the French National Institute for Agricultural Research on behalf of ADEME[8] (De Cara et al., 2012), which provided a critical review of evaluations of the effect of LUC on the overall environmental impacts of biofuels, found that "Almost 90% of the collected evaluations conclude that the development of biofuels leads to (direct or indirect) LUC that cause GHG emissions." However, the debate continues over whether the scope of studies on the carbon footprint of biofuels should expand to account for LUC. Defining emission factors for both components separately makes it possible to assess the complete potential climate impacts of biofuel consumption, while identifying the different underlying drivers.

2.2 Emission factors

In this section, distinct emission factors are estimated for each category of biofuels—and associated feedstocks—considered, which are listed in Table 8.2. Overall, 13 types were distinguished for gasoline substitutes (ethanol) and 10 for diesel substitutes (biodiesel and renewable diesel).

---

[6] The Fifth Assessment Report of the IPCC defines an emission factor as "the emissions released per unit of activity" (Allwood et al., 2014). Similarly, the United Nations Framework Convention on Climate Change (UNFCCC) defines it as "the average emission rate of a given GHG for a given source, relative to units of activity" (UNFCCC GHG data—UNFCCC, Definitions). The units of activity we consider in this study correspond to the consumption of a certain amount of biofuel, expressed in volume or in energy content.

[7] CO_2-equivalent emissions are "The amount of carbon dioxide (CO_2) emissions that would cause the same integrated radiative forcing, over a given time horizon, as an emitted amount of a greenhouse gas (GHG) or a mixture of GHGs. [...] CO_2-equivalent emission is a common scale for comparing emissions of different GHGs but does not imply equivalence of the corresponding climate change responses" (Allwood et al., 2014).

[8] The French Environment and Energy Management Agency.

As mentioned in the previous section, the overall "CO_2e performance" of biofuels can be split into two components: the carbon dioxide emitted along the value chain, and the emissions related to changes in land use. Three coefficients are hence reported in Table 8.2:

- a "GHGWTW" component, corresponding to well-to-wheels[9] emissions (upstream processes + combustion);
- a "GHGLUC" component, corresponding to emissions from direct and indirect LUC;
- "GHGTOT": the total CO_2e performance (= GHGWTW + GHGLUC).

The emission factors obtained are not differentiated according to the country of production. In the next section, we will input these estimates within a macroeconomic partial equilibrium model to assess the aggregated contribution of biofuels to emission reductions up to 2017, and these estimates will be applied similarly for all countries. Two main arguments support this choice. First, the value of emission factors varies widely depending on the types of models, methods that were used, and assumptions made (for example, regarding the modeling of markets, the representation of land use, etc.). To have a coherent dataset, we decided to restrict the number of studies reviewed to focus on those that were most recent and had a relatively wide scope. This prevents us from comparing data obtained from contradictory hypotheses. Second, in the context of macroeconomic modeling research, it is neither practical nor efficient to use such a complex set of inputs.

It is worth noting that using the same emission factor regardless of the specificities of the country of production has undeniable limits. Let us look at a nontrivial example: the case of sugarcane-based ethanol. In the "two degrees" scenario of the International Energy Agency (IEA)'s 2016 Energy Technology Perspectives report (IEA, 2016) as well as in the US legislation, a distinction is made between sugarcane-based ethanol and other conventional ethanol feedstocks, due to the former being considered responsible for relatively fewer GHG emissions. However, the alleged low carbon intensity of sugarcane ethanol relies on the use of sugarcane bagasse as the source of heat in the conversion process (a carbon neutral source), instead of the usual coal. While this is indeed the technology used in Brazil, other countries might use different sources for heat. The numbers in Table 8.1 are calculated from a study that was conducted in 2009 to inform the development of the UK's Renewable Transport Fuel Obligations legislation (Renewable Fuels Agency, 2009). These results show that relying on certain energy sources (such as coal) for the biofuel conversion process increases the expected amount of emissions by a factor of about 4.

Because the modeling exercise will focus on the major biofuel-consuming countries, and because the immense majority of sugarcane-based ethanol is produced in Brazil, this consideration is not necessary for our scope of analysis. However, the foregoing example

---

[9] We use the term "well-to-wheels" (WTW) for transport fuels to represent the steps the product goes through from its extraction to its combustion in the vehicle engine. WTW emissions are therefore the sum of well-to-tank (WTT) and tank-to-wheels (TTW) emissions, where:
- WTT refers to the cascade of steps required to produce and distribute a fuel (starting from the primary energy resource), including vehicle refueling;
- TTW corresponds to the combustion of the fuel in the vehicle.

Table 8.1 Comparison of the carbon intensity along the production process of sugarcane ethanol in several countries.

Stage/country	Brazil	Mozambique	Pakistan	South Africa
Emissions	Unit: (kg CO_2e/t ethanol)			
1. Crop production	348	425	597	425
2. Feedstock transport	49	53	49	53
3. Conversion	**0**	**0**	**2152**	**2219**
4. Liquid fuel transport	268	338	296	328
Total emissions				
kg CO_2e/t biofuel	665	816	3094	3025
kgCO_2e/GJ	24.7	30.3	114.9	112.4
kgCO_2e/m^3	**519**	**637**	**2413**	**2360**

From Renewable Fuels Agency. Carbon and Sustainability Reporting Within the Renewable Transport Fuel Obligation, Technical Guidance Part Two Carbon Reporting—Default Values and Fuel Chains, 2009.

illustrates that sugarcane-based ethanol is not automatically a "low-carbon" fuel. As aforementioned and discussed also in Chapter 2, the assessment of the carbon intensity of biofuels should consider the whole conversion process in the biofuel pathway rather than solely the commodity of origin.

The two main studies considered are, for WTW emissions, the study conducted in 2014 by the JEC Consortium (Edwards et al., 2014), and for LUC emissions, the study commissioned by the European Commission and conducted by the Ecofys, IIASA, and E4tech consortium (Valin et al., 2015) in 2015, which is based on the GLOBIOM economic model. For certain categories of biofuels, emission factors were estimated by calculations based on more specific studies. The results of Hoefnagels et al. (2010) and Overmars et al. (2015) were used to generate estimates for dedicated energy crops (in particular herbaceous perennials) and biodiesel produced from jatropha. The studies by Koga and Tajima (2011) and Nguyen et al. (2007) provided evaluations of WTW emissions for rice and cassava (a tuber), respectively.

Table 8.2 and Fig. 8.1 present the estimates thereby constructed and retained to evaluate the carbon intensity of the main categories of biofuels.

In the rest of this chapter, we will use the general term "biodiesel" to refer to both fatty acid methyl esters (the generic chemical term for biodiesel) and renewable diesel (such as Fischer-Tropsch diesel).[10]

Fig. 8.1 distinctly shows the hierarchy between conventional ethanol and biodiesel, in terms of the amount of GHG emissions induced. We can see that the different categories

---

[10] Fatty acid ethyl esters also fall under the "biodiesel" category, while renewable diesel also includes other products such as hydrogenated esters and fatty acids and hydrogenated pyrolysis oil. Figs. 8.4 and 8.5 in Chapter 2 provide a detailed understanding of the multiple biofuels categories and their production pathways.

Table 8.2 Carbon intensity of the main categories of biofuels ($kgCO_2e/m^3$).

Final fuel	Feedstock	GHG-WTW	GHG-LUC	GHG-TOT
Ethanol	Sugarcane	521	357	878
Ethanol	Sugar beet	846	315	1161
Ethanol	Molasses	684	336	1020
Ethanol	Corn	1686	294	1980
Ethanol	Other coarse grains	1596	756	2352
Ethanol	Wheat	1457	714	2171
Ethanol	Rice	1391	487	1879
Ethanol	Roots and tubers	964	487	1451
Ethanol	Agricultural residues	193	336	529
Ethanol	Forestry residues	410	357	767
Ethanol	Dedicated energy crops	459	−294	165
Biodiesel	Palm oil	1664	7739	9403
Biodiesel	Soybean oil	1846	5025	6871
Biodiesel	Other vegetable oil	1631	2144	3775
Biodiesel	Jatropha	1419	4290	5709
Biodiesel	Waste oils and fats[a]	608	0	608
Fischer–Tropsch diesel	Agricultural residues	85	544	629
Fischer–Tropsch diesel	Forestry residues	85	578	663
Fischer–Tropsch diesel	Dedicated energy crops	325	−697	−372

[a]Waste oils and fats include used cooking oil and rendered animal fats (such as tallow, lard, and chicken fat).

of conventional biomass-based diesel induce more than twice the amount of emissions related to the production of ethanol. LUCs are responsible for the majority of these emissions. This can notably be explained by the type of land that is often affected, namely peatlands. The expansion of palm tree cultivation for oil especially leads to the drainage of peatlands. As peat is oxidized, significant amounts of carbon dioxide are released into the atmosphere.

Furthermore, biomass-based diesel derived from waste oils and fats is shown to have the lowest carbon intensity of all categories of biodiesel that have reached technological maturity. Because the original feedstock was not produced with the intention of converting it into biofuel, potential LUC impacts of the production of that material (before it was used and turned into waste) should arguably not be ascribed to the end-fuel. This assumption, however, can be challenged.

Fig. 8.1 further shows the low-carbon intensity of advanced biofuels—in comparison with conventional biofuels—the three types of feedstocks considered being agricultural residues, forestry residues, and in the case of biodiesel, waste oils and fats.

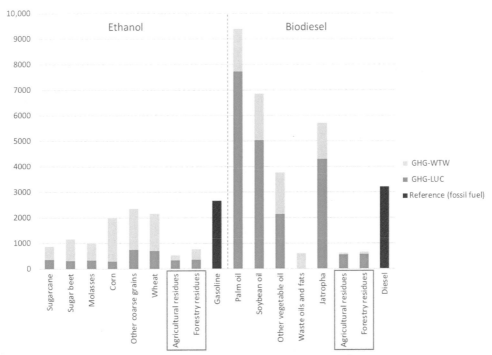

Fig. 8.1 Carbon intensity of different categories of biofuels (kgCO$_2$e/m^3). *GHG*, greenhouse gas; *LUC*, land use change; *WTW*, well-to-wheels.

2.3 From absolute emissions to emissions savings

GHG emissions savings from the use of biofuels refer to the difference between the amounts of GHG actually emitted by the use of biofuels and the emissions that would have arisen from an equivalent amount of energy being supplied by fossil fuels. They are therefore computed by comparing—on an energy content basis—the carbon intensity of ethanol to that of gasoline and the carbon intensity of biodiesel to that of diesel, in kgCO$_2$e/MJ. The difference between the two is then applied to the volume of energy that was substituted to obtain the total quantity of emissions that were prevented. Table 8.4 summarizes the energy content by volume of each type of fuel.

Estimations of the carbon intensity of fossil fuels should account for the totality of emissions embodied in the product, taking into account both upstream emissions (related to the extraction, transport, refining, and distribution processes) and emissions stemming from the actual combustion of the fuel. Table 8.3 details the contribution of each component to the total carbon intensities of gasoline and diesel.

Table 8.3 Emission factor and energy content by volume of transport fuels

Measure	Unit	Fuel		Source
		Gasoline	Diesel	
Upstream emissions	$kgCO_2e$/GJ LHV	13.8	15.4	Edwards et al. (2013)
Combustion emissions	$kgCO_2e$/GJ LHV	69.3	74.1	IEA (2017a,b)
Total emissions	$kgCO_2e$/GJ LHV	83.1	89.5	–
LHV[a]	GJ/t	43.2	43.1	Edwards et al. (2013)
Density	kg/m^3	745	832	Edwards et al. (2013)
LHV	GJ/m^3	32.2	35.9	–
Total emissions	$kgCO_2e/m^3$	2674	3208	–

[a] The lower heating value (LHV) of a fuel, also known as net calorific value, is a measure of the amount of heat produced by a complete combustion of that fuel.

3. Assessing the GHG emissions savings from biofuel use since 2000

The estimations presented in this section of GHG emissions savings derived from the use of biofuels in the transport sector, both historical and forecasted, are based on the Aglink-Cosimo model.

3.1 Aglink-Cosimo: A partial equilibrium model of world agricultural markets

Aglink–Cosimo is a macroeconomic model of world agricultural markets, developed jointly by the Organization for Economic Co-operation and Development (OECD) and the Food and Agriculture Organization (FAO). It is a partial equilibrium model, i.e., markets outside of the agricultural scope are not explicitly represented and major macroeconomic variables—such as exchange rates, GDPs, and population growth—are entered as exogenous inputs. The model distinguishes 56 countries or regions. The biofuel component of the Aglink-Cosimo model is described in detail in OECD (2018).

The corresponding database is updated yearly to account for new policies in place and new trends in agricultural markets. This update is nourished by the expertise of economists at the OECD and FAO and questionnaires sent to member countries. The output of the simulation of the updated model is twofold:

1. a historic view of market variables up to the present year;
2. market projections for the next 10 years, corresponding to a BAU scenario, i.e., what can be expected if current policy settings continue into the future.

For the purpose of this analysis, we focus on the major biofuel consumer countries or groups of countries. We consider 10 countries for biodiesel and 10 for ethanol, each set representing more than 97% of global consumption in 2017. Fig. 8.2 shows the evolution of the global volumes consumed and the distribution among these key countries since 2000.

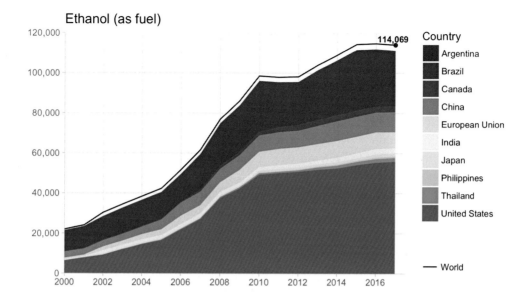

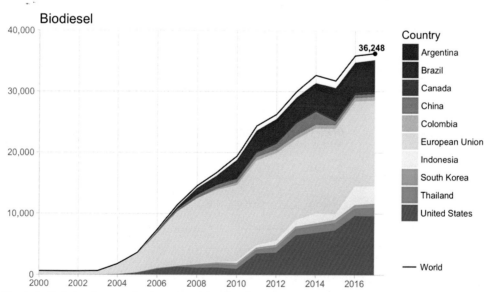

Fig. 8.2 Global volumes of ethanol and biodiesel consumed and their distribution among major consumer countries (in mln L). *(From OECD-FAO, 2018. OECD-FAO Agricultural Outlook Database accessed online—Available from: https://stats.oecd.org/Index.aspx?datasetcode¼HIGH_AGLINK_2018).*

3.2 Calculating emissions savings

For ethanol, total GHG emissions savings are calculated as the ratio of the difference between the total emissions associated with ethanol use and the total "reference" emissions that would have occurred if gasoline had been used instead, over the total emissions associated with gasoline use in the given year. Savings are computed similarly for biodiesel, with diesel as the corresponding reference fossil fuel. Box 8.1 reiterates and details further the steps performed to construct the final emissions savings estimates.

As described, the aggregate carbon intensity of a biofuel is the sum of the GHGs emitted along the transformation and combustion process (WTW emissions) and those arising from LUC (LUC emissions). One can therefore decompose the total emissions savings between the savings realized along the value chain and those related to LUC. However, as fossil fuels do not induce LUCs (at least, arguably, not of any order of magnitude of interest here), the *GHG_LUC* reference term for fossil fuels is equal to zero. As a result, any LUC induced by the production of biofuels generates *negative* GHG savings, i.e., emissions that would not have taken place had fossil fuel been used.

BOX 8.1 Detail of the calculation of emission savings from biofuel use; example of ethanol

For a given country in a given year, we define:

V_{e_f} the volume of ethanol consumed that was produced from feedstock of type f

$\varepsilon_{gas,e}$ the energy content ratio between ethanol and gasoline: $\varepsilon_{gas,e} = \frac{[GJ/m^3 ethanol]}{[GJ/m^3 gasoline.]}$

$V_{gas,sub}$ the volume of gasoline substituted by ethanol: $V_{gas,\,sub} = V_{e_f} \times \varepsilon_{gas,e}$

$V_{gas,cons}$ the volume of gasoline consumed

Emissions (in $ktCO_2e$):

$$\overline{GHG\_WTW}_e = \sum_f GHG\_WTW_{e_f} \times V_{e_f}$$

$$\overline{GHG\_LUC}_e = \sum_f GHG\_LUC_{e_f} \times V_{e_f}$$

$$\overline{GHG\_TOT}_e = \sum_f GHG\_TOT_{e_f} \times V_{e_f} = \overline{GHG\_WTW}_e + \overline{GHG\_LUC}_e$$

$$\overline{GHG}_{gas,sub} = GHG_{gas} \times V_{gas,sub}$$

$$\overline{GHG}_{gas,cons} = GHG_{gas} \times V_{gas,cons}$$

Emissions savings (in %):

$$WTW\ GHG\ savings = \frac{\overline{GHG\_WTW}_e - \overline{GHG}_{gas,sub}}{\overline{GHG}_{gas,cons}}$$

$$LUC\ GHG\ savings = \frac{\overline{GHG\_LUC}_e - 0}{\overline{GHG}_{gas,cons}}$$

$$Total\ GHG\ savings = \frac{\overline{GHG\_TOT}_e - \overline{GHG}_{gas,sub}}{\overline{GHG}_{gas,cons}} = WTW\ GHG\ savings + LUC\ GHG\ savings$$

The relationships just described for the example of ethanol lead to the following property of the overall mitigation performance of biofuels:

$$Total\ GHG\ savings = WTW\ GHG\ savings + LUC\ GHG\ savings$$

Estimates of the total emissions savings achieved by substituting fossil fuels with biofuels will be sensitive to the inclusion of direct and indirect LUC emissions. Separating the WTW and LUC components enables us to disentangle the respective impacts of these two underlying drivers and highlight their respective orders of magnitude. This is especially relevant because the debate on whether estimations of the carbon footprint of biofuels should consider emissions related to LUC remains open.

3.3 Results: Historical emissions savings until 2017 and medium-term projections in a BAU scenario

Figs. 8.4 and 8.5 present the main results from the modeling exercise for ethanol and biodiesel, respectively. In each figure, the historical annual total emissions savings and their projected levels up to 2027[11] are presented in comparison with the levels of biofuel incorporation (in both volume and energy terms). The decomposition into WTW- and LUC-related savings is also provided for the latest year corresponding to historical data, namely 2017.

The following results correspond to the volumes consumed globally. However, this exercise could also be performed at the country level, by allocating the emissions to the biofuel consumer countries. This analysis would require some additional steps. Indeed, Aglink-Cosimo is constructed such that different feedstocks of origin are represented on the supply side, while on the demand side only the two "ethanol" and "biodiesel" aggregates remain. Moreover, the model does not contain a spatial representation of trade; it assumes homogeneity on the world market for all commodities, i.e., importers do not distinguish commodities by country of origin (OECD, 2015). To attribute the GHG emissions to the consumer country, one must account for the characteristics of the biofuel considered, which includes the feedstock of origin. Conducting this exercise would therefore entail estimating trade fluxes and reallocating biofuels exchanges by country—and thereby feedstocks—of origin, while ensuring internal consistency of volumes traded.

Historical emissions savings

Consumption of ethanol has led to annual total emissions savings that increased slightly from 0.3% in 2002 to about 0.5% in 2017, while the average volumetric blending rate has increased from around 2% at the beginning of the century to over 7.6% in 2017. This relatively low mitigation performance is notably explained by emissions arising

---

[11] The projections correspond to a BAU scenario, i.e., what can be expected if current policy settings continue into the future.

Table 8.4 Energy content by volume of transport fuels (in GJ/m^3)

Fuel	Gasoline	Diesel	Ethanol	Biodiesel
Energy content	32.2	35.9	21	33

From Edwards, R., Larivé, J.-F., Rickeard, D., Weindorf., 2014 Well-to-Wheels analysis of future automotive fuels and powertrains in the European context, WELL-TO-TANK (WTT) Report Version 4a. Technical report, European Commission, EUR 26237—Joint Research Centre—Institute for Energy and Transport; Directive 2009/28/EC of the European Parliament and of the Council of 23 April 2009 on the promotion of the use of energy from renewable sources and amending and subsequently repealing Directives 2001/77/EC and 2003/30/EC (Text with EEA relevance) - ANNEX III, Energy content of transport fuels.

from LUC, which partially counteract the savings achieved along the value chain, as underlined by panel b in Fig. 8.4. However, even when restricting the scope to WTW emissions, savings remain well below the incorporation rate. Two main factors drive this result: we saw in the first section of this chapter that the production of biofuels still generates GHG emissions at each step of the transformation and transport process.[12] Second, the incorporation rate displayed in Fig. 8.4 as a solid line corresponds to the volumetric blending rate. Because ethanol has a lower energy content than gasoline (Table 8.4), a higher volume of ethanol is needed to supply the same amount of total energy than would gasoline. Ethanol produced from grains (corn, barley, rye, or wheat) is therefore more carbon intensive than gasoline on an energy content basis (Fig. 8.1). Finally, the mix of feedstocks in the global volumes varied slightly across the period considered, as shown in Fig. 8.3, which explains the small changes in the ratio of the blending rate and total emissions savings.

The case of biodiesel paints a worse picture (Fig. 8.5), because emissions from LUCs completely offset the mitigation gains from burning biomass-based fuel instead of fossil fuel. In 2017, global emissions savings were therefore negative, estimated at around −3.0%, i.e., the use of biodiesel resulted overall in higher emissions than if solely diesel had been used. Across the period, these negative savings were also higher than the incorporation rate. However, we can see that when considering only WTW emissions, the consumption of biodiesel did generate positive emission savings.

Medium-term projections

If current policies remain as they are, and production remains dominated by conventional biofuels, projections over the next 10 years suggest virtually no increase in mitigation levels.

These results should be analyzed in parallel with the carbon intensities of the different biofuel categories, displayed in Fig. 8.1. Three categories stood out as having a much higher

[12] Because the amount of carbon released by the combustion of biofuels is equal to the amount initially absorbed by the plant, the combustion stage is deemed carbon neutral; however, the transport and conversion of the initial feedstock into the final biofuel product engender nonnegligible levels of emissions.

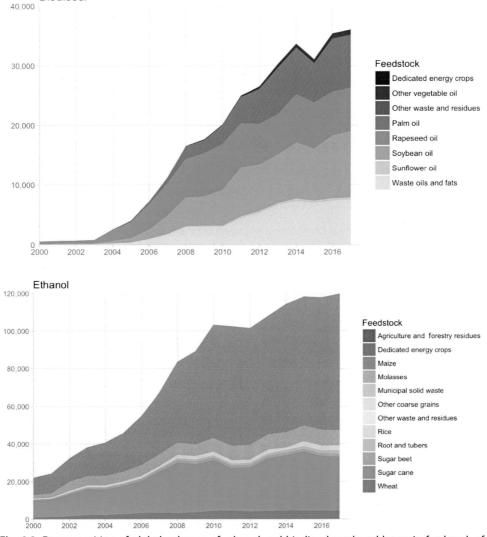

Fig. 8.3 Decomposition of global volumes of ethanol and biodiesel produced by main feedstock of origin (in mln L). *(From OECD-FAO, 2018. OECD-FAO Agricultural Outlook Database accessed online—Available from: https://stats.oecd.org/Index.aspx?datasetcode¼HIGH_AGLINK_2018).*

mitigation potential than their conventional counterparts, namely agricultural residues, forestry residues, and in the case of biodiesel, waste oils and fats. If the composition of current volumes produced were to change in favor of these *advanced* categories, the mitigation performance of biofuels could be increased substantially.

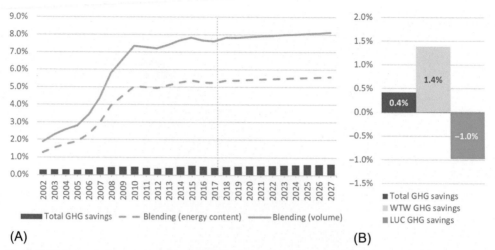

Fig. 8.4 Global emissions savings from ethanol use over the period 2002–27. (A) Evolution of the total savings in relation to the average incorporation rate; the *vertical dotted line* separates the historical data (*left*) from the business-as-usual scenario (*right*). (B) Decomposition of the total greenhouse gas savings in its well-to-wheels and land use change-related savings components in 2017. *GHG*, greenhouse gas; *LUC*, land use change; *WTW*, well-to-wheels.

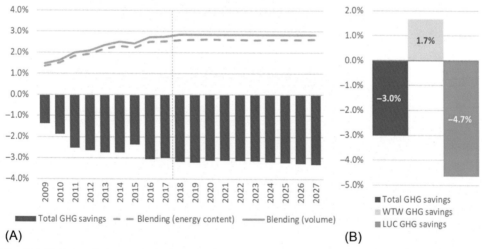

Fig. 8.5 Global emissions savings from biodiesel use over the period 2009–27. (A) Evolution of the total savings in relation to the average incorporation rate; the *vertical dotted line* separates the historical data (*left*) from the business-as-usual scenario (*right*). (B) Decomposition of the total greenhouse gas savings in its well-to-wheels and land use change-related savings components in 2017. *GHG*, greenhouse gas; *LUC*, land use change; *WTW*, well-to-wheels.

4. Conclusion

On a global scale, when both WTW and LUC emissions are taken into account, biofuel markets so far seem to have induced limited or zero emission savings in the transportation fuel sector. The medium-term perspective is similar if policies—and available technologies—remain as they are today.

This assessment underlines the necessity to switch to less carbon-intensive types of biofuels if they are to contribute to climate-change mitigation in the transport sector. However, the extent to which biofuels could contribute to lower the carbon intensity of this sector highly depends on ongoing technological developments and the policy environment. The next decade could be determinant in that regard: the technologies that will help us achieve long-term emission reduction targets are those that are being developed today at a laboratory or prototype scale.

Chapter 14 focuses on possible evolutions by 2030. The Aglink-Cosimo model is used to conduct simulations based on the specific emissions targets of the IEA's 2 degrees scenario (2DS). By explicitly considering the interconnection between agricultural and biofuel markets, we explore the medium-term ability of the agricultural sector to supply the amount of bioenergy deemed necessary to meet mitigation targets in the transport sector. In particular, given the policies and intrinsic dynamics of the agricultural sector, we analyze whether the development of biofuels as envisioned in the 2DS is indeed achievable with the incentives considered (notably, increasing carbon tax levels on transportation fuels).

Disclaimer

The views expressed are those of the author and do not necessarily represent the official views of the OECD or of the governments of its member countries.

References

Allwood, J.M., Bosetti, V., Dubash, N.K., Gómez-Echeverri, L., von Stechow, C., 2014. Glossary. In: Edenhofer, O., Pichs-Madruga, R., Sokona, Y., Farahani, E., Kadner, S., Seyboth, K., Adler, A., Baum, I., Brunner, S., Eickemeier, P., Kriemann, B., Savolainen, J., Schlömer, S., von Stechow, C., Zwickel, T., Minx, J.C. (Eds.), Climate Change 2014: Mitigation of Climate Change. Contribution of Working Group III to the Fifth Assessment Report of the Intergovernmental Panel on Climate Change. Cambridge University Press, Cambridge, UK; New York, NY.

De Cara, S., Goussebale, A., Grateau, R., Levert, F., Quemener, J., Vermont, B., Bureau, J.-C., Gabrielle, B., Gohin, A., Bispo, A., 2012. Revue critique des études évaluant l'effet des changements d'affectation des sols sur les bilans environnementaux des biocarburants. Rapport de l'ADEME.

Edwards, R., Larivé, J.F., Rickeard, D., Weindorf, W., 2013. Well-to-Wheels Analysis of Future Automotive Fuels and Powertrains in the European Context, Well-to-Tank Appendix 4—Version 4.0, Description, Results and Input Data per Pathway. Technical report, European Commission, EUR 26028—Joint Research Centre—Institute for Energy and Transport.

Edwards, R., Larivé, J.-F., Rickeard, D., Weindorf, W., 2014. Well-to-wheels analysis of future automotive fuels and powertrains in the European context. In: Well-to-Tank (WTT) Report Version 4a. Technical

Report. European Commission, EUR 26237—Joint Research Centre—Institute for Energy and Transport.

Hoefnagels, R., Smeets, E., Faaij, A., 2010. Greenhouse gas footprints of different biofuel production systems. Renew. Sust. Energ. Rev. 14 (7), 1661–1694.

International Energy Agency, 2016. Energy Technology Perspectives report 2016: Towards Sustainable Urban Energy Systems. OECD.

International Energy Agency, 2017a. Energy Technology Perspectives report 2017: Catalysing Energy Technology Transformations. OECD.

International Energy Agency, 2017b. World CO2 Emissions From Fuel Combustion: Database Documentation, 2017 edition. IEA.

Khanna, M., Crago, C.L., 2012. Measuring indirect land use change with biofuels: implications for policy. Ann. Rev. Resour. Econ. 4 (1), 161–184.

Koga, N., Tajima, R., 2011. Assessing energy efficiencies and greenhouse gas emissions under bioethanol-oriented paddy rice production in northern Japan. J. Environ. Manag. 92 (3), 967–973.

Nguyen, T.L.T., Gheewala, S.H., Garivait, S., 2007. Energy balance and GHG abatement cost of cassava utilization for fuel ethanol in Thailand. Energy Policy 35 (9), 4585–4596.

OECD, 2015. Aglink-Cosimo Model Documentation—A partial equilibrium model of world agricultural markets. Technical report, OECD.

OECD, 2018. Biofuel module documentation. Accessed online, http://www.agri-outlook.org/about/Aglink-Cosimo-Biofuel-Documentation.pdf. (Accessed April 2018).

Overmars, K., Edwards, R., Padella, M., Prins, A.G., Marelli, L., Consultancy, K.O., 2015. Estimates of Indirect Land Use Change From Biofuels Based on Historical Data. Publications Office of the European Union, Luxembourg.

Renewable Fuels Agency, 2009. Carbon and Sustainability Reporting Within the Renewable Transport Fuel Obligation, Technical Guidance Part Two Carbon Reporting—Default Values and Fuel Chains.

Valin, H., Peters, D., van den Berg, M., Frank, S., Havlik, P., Forsell, N., Hamelinck, C., Pirker, J., Mosnier, A., Balkovic, J., Schmidt, E., Durauer, M., Di Fulvio, F., 2015. The Land Use Change Impact of Biofuels Consumed in the EU: Quantification of Area and Greenhouse Gas Impacts. European Commission

Further reading

Directive 2009/28/EC, 2009. Directive 2009/28/EC of the European Parliament and of the Council of 23 April 2009 on the promotion of the use of energy from renewable sources and amending and subsequently repealing Directives 2001/77/EC and 2003/30/EC (Text with EEA relevance) - ANNEX III, Energy content of transport fuels.

OECD-FAO, 2018. OECD-FAO Agricultural Outlook Database. Available from: https://stats.oecd.org/Index.aspx?datasetcode=HIGH_AGLINK_2018.

United Nations Framework Convention on Climate Change, GHG Data—UNFCCC, Definitions. http://unfccc.int/ghg_data/online_help/definitions/items/3817.php. (Accessed 1 August 2016).

SECTION 4

Analyzing policy options

CHAPTER 9

Consequences of US and EU biodiesel policies on global food security

Deepayan Debnath*, Jarrett Whistance*, Patrick Westhoff*, Mike Helmar[†]
*Food and Agricultural Policy Research Institute, University of Missouri, Columbia, MO, United States
[†]University of Nevada-Reno, Reno, NV, United States

Contents

1. Introduction

There is no doubt that government policies have encouraged both the production and use of ethanol and biodiesel over the past decade. From 2010 to 2017, world ethanol production increased by 25%, while during the same period biodiesel production increased by 158% (F.O. Lichts, 2018). At the same time, those policies faced scrutiny for their perceived role in the 2007–08 food price spikes (FAO, 2008; Babcock, 2011; Wright, 2014; Rosegrant, 2008; Hochman et al., 2010). Naylor and Higgins (2018) pointed out that, after the 2007–08 price spikes, the real vegetable oil price went back to the early 2000s price path, and the authors expected the downward pressure on the vegetable oil price to continue. They also highlighted the potential food security implications of increased global biodiesel production.

Vegetable oil consumption is increasing in many developing countries, and could account for 13% of global calorie supplies in 2050 according to FAO (2012). On the other hand, economic growth in those countries leads to greater demand for diesel-powered trucking services and expansion of their public transportation sector, which mostly relies on diesel-engine buses. In brief, the demand for food and fuel rises simultaneously in developing countries. Policymakers face the challenge of addressing the food-versus-fuel debate. In this study, we address some broader questions: What are

Biofuels, bioenergy and food security
https://doi.org/10.1016/B978-0-12-803954-0.00009-7

the consequences of certain biodiesel mandates on agricultural commodity prices? We provide further evidence of the relationship between the biodiesel and agricultural markets and trace the consequences of those biodiesel mandates on grains and vegetable oil prices. Unlike previous studies focusing on ethanol, we shed light on the impact of key US and EU biodiesel policies on the grain and vegetable oil markets using a multimarket, multiregion dynamic partial-equilibrium economic model.

The links between food and biofuel markets are influenced by biofuel policies. In 2009, the European Union targeted a renewable fuel rate of 10% in the total transportation energy consumed by the member countries by 2020, while the use of biofuels derived from food-based feedstocks is capped at 7% (EU Commission, 2015). The consumption of biofuel in the United States is driven in large part by the Renewable Fuel Standard (RFS) established by the 2007 Energy Independence and Security Act, which targeted the use of 36 billion gallons of biofuel by the end of 2022 (EISA, 2007). Of that total, there are three submandates: one for advanced biofuels that meet at least a 50% reduction in greenhouse gas (GHG) emissions; another for biomass-based diesel, which also must meet the 50% reduction in GHG emissions; and a requirement for cellulosic fuels that must meet a 60% reduction in GHG emissions. Because of the hierarchical nature of these requirements, fuels that meet the submandates can also qualify for the broader mandates. For example, biomass-based diesel can be used to help meet the overall requirements, in addition to conventional ethanol, if market conditions warrant.

While total biofuel use may never reach the original targets, its use is far greater than it would have been in the absence of supportive policies. Countries with the fastest growing economies, including India and China, have also targeted biofuel blending with traditional transportation fuel (USDA-FAS GAIN Reports, 2015a,b). The governments of developing countries have also shown interest in introducing biofuel to reduce their dependency on imported fossil fuel (Müller et al., 2008; Pingali et al., 2008; Ewing and Msangi, 2009). However, before implementing any biofuel policy, local governments might need to answer the question of whether or not the introduction of a new biofuel policy could increase food and feed prices.

2. Policies targeting biodiesel and its interaction with the agricultural market

In the United States, biodiesel use has increased significantly because of increases over time in the RFS mandates for biodiesel and advanced biofuels (EISA, 2007; EPA, 2017). The use of biodiesel is even greater in the European Union, largely because of policies to encourage or require its use. The European Commission encourages the use of nonfood-based feedstocks in biodiesel production because they generate higher computed GHG savings compared to the production of biodiesel from vegetable oil (EU Commission, 2015). In this study, the impacts of policies in major biodiesel-producing countries are analyzed and their consequences on the world agricultural market are discussed.

Serra et al. (2011a,b), Thompson et al. (2010, 2011), Thompson and Meyer (2013), Whistance and Thompson (2014), Timilsina, and Shrestha (2011), Zhang et al. (2010), and Hochman et al. (2014) analyze the links between the biofuel, fuel, and crop markets. The broad consensus is that an increase in energy prices that leads to a higher ethanol price can, consequently, lead to an increase in the prices of feedstocks. Rosegrant (2008) used a simulation-based model to estimate the effect of biofuels on food crop prices. He concluded that, with the growth in demand for biofuel between 2000 and 2007, the weighted average grain price was increased by 30%. He found the food crop price increase was highest for maize (39%) but not as large for wheat (22%) and rice (21%). The complexity of the price dynamics of different food crops was described by Chen et al. (2011) and Chakravorty, Hubert, and Nøstbakken (2009). In another study, Abbott, Hurt, and Tyner (2008) reviewed several studies on the global food crisis and conclude that, instead of any particular cause, there were several factors that resulted in food price inflation, including depreciation of the US dollar, global changes in weather patterns, changes in food consumption patterns in developing countries, and the introduction of biofuel policies by the US and EU governments. However, these studies did not address the future consequences of an increase or decrease in agricultural commodity prices, including those for grains and oilseeds, due to certain biodiesel-related policy changes. Our study traces both the direct and indirect impacts of recent biodiesel policy developments on agricultural markets with regards to the changes in agricultural commodity prices.

Increasing the use of vegetable oil for the production of biodiesel could result in upward pressure on vegetable oil prices, while increasing the share of biodiesel produced from nonfood-based feedstocks could lead to downward pressure on vegetable oil prices. When changes in biodiesel production cause changes in the production of oilseeds like soybeans and rapeseed, the result is a change in the availability and price of oilseed meals used as livestock feed. To estimate the consequences of biodiesel policies for vegetable oil, grain, and feed markets, we use a system of commodity models that incorporates important cross-country and cross-commodity linkages. By doing so, we differentiate this research study from the existing literature on biodiesel policies. The objectives of this study are to determine the consequences of biodiesel mandates introduced by major biofuel-producing and -consuming countries, specifically the United States and the European Union, on global grain, vegetable oil, and livestock feed prices.

3. International biodiesel model

A structural economic model is developed and used to represent the international biodiesel market that interacts with the international agricultural commodity and livestock markets to simulate the consequences of country-specific biofuel policies on the overall biodiesel and biodiesel feedstock markets (Barr et al., 2011; Dumortier et al., 2011; Meyer and Thompson, 2012; Thompson and Meyer, 2013;

Egbendewe-Mondzozo et al., 2015; Thompson et al., 2010; Debnath et al., 2017a,b). The biodiesel component is part of an international biofuels model that is a multimarket, multiregion, dynamic, partial-equilibrium model consistent with the approach used in OECD-FAO and USDA-ERS agricultural modeling systems (OECD-FAO, 2017; USDA-OCE, 2018). It includes the major world biodiesel-producing and -consuming countries. On the demand side, we consider domestic use, exports, and ending stocks, while the supply side consists of beginning stocks, production, and imports. It is consistent across the countries that are represented.

In this study we consider the major food-based feedstocks used to produce biodiesel, which are soy oil, rapeseed oil, and palm oil. Different countries rely on different feedstocks, and we include multiple feedstocks in countries where appropriate. For example, biodiesel produced in the United States, Brazil, and Argentina is derived mainly from soy oil, though rapeseed oil and corn oil might also be used. Moreover, roughly half of the total biodiesel produced in the EU is derived from rapeseed oil, and palm oil is the major feedstock used to produce biodiesel in Indonesia and Malaysia. The international biodiesel model solves for the equilibrium biodiesel price and quantity for each year taking into account the existing mandates, tax credits, and trade policies for each (country-specific) biofuel market. It also generates the demand for feedstocks required in the production of biodiesel, which drives the vegetable oil prices, because it is linked to the international agricultural commodity market model. The use of a structural economic model allows us to explore the impacts of changes in the biofuel policies on biodiesel and feedstock markets in the United States, the European Union, and around the world.

Unlike ethanol, where only low-level blends can be used by most conventional gasoline-powered engines, most diesel combustion vehicles can easily switch between diesel and biodiesel blends, so discretionary demand for biodiesel is triggered primarily by relative prices of biodiesel and petroleum-based diesel fuel (USDA-FAS, 2015a,b). Beyond modeling country-specific biodiesel use driven by the mandates, we consider the substitution effect between biodiesel and diesel while modeling the biodiesel market (Fig. 9.1). Discretionary biodiesel use in each country is modeled as a function of biodiesel (own) and diesel (cross) prices and income. The following equation explains biodiesel demand:

$$Q_t^{BD} = \alpha_0 + \alpha_1 \frac{BD_t^P}{DI_I^P} + \alpha_2 I_t + \alpha_3 \max\left(0, \left(0.90 - \frac{BD_t^P}{DI_t^P}\right)\right)$$
$$+ \alpha_4 \max\left(0, \left(0.87 - \frac{BD_t^P}{DI_t^P}\right)\right) + \alpha_5 \max\left(0, \left(0.84 - \frac{BD_t^P}{DI_t^P}\right)\right) + \varepsilon_t \tag{9.1}$$

where Q_t^{BD} is the per capita biodiesel use for a particular country; I_t is the per capita GDP; BD_t^P and DI_t^P are the biodiesel and diesel prices in year t; α_0, α_1, α_2, α_3, α_4, and α_5 are intercept and slope coefficients; and ε_t is the corresponding error term. The max function

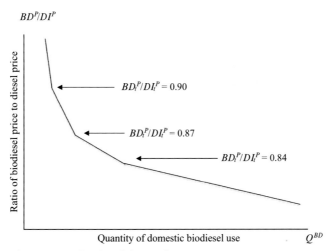

BD^P/DI^P

Ratio of biodiesel price to diesel price

$BD_t^P/DI_t^P = 0.90$

$BD_t^P/DI_t^P = 0.87$

$BD_t^P/DI_t^P = 0.84$

Quantity of domestic biodiesel use Q^{BD}

Fig. 9.1 Modeling discretionary biodiesel demand.

is introduced to capture the asymmetric increase in biodiesel demand due to changes in the relative price of biodiesel to diesel. The energy equivalent coefficient between biodiesel and diesel is 0.90 (US EIA, 2018), so we expect consumer response to be more elastic once this threshold is met. This response is crucial in capturing the discretionary demand for biodiesel. In the case where rising global crude prices lead to increases in the price of diesel, such that $\frac{BD_t^P}{DI_t^P} < 0.87$, then the associated terms (α_3 and α_4) will be triggered, and when DI_t^P further increases resulting in $\frac{BD_t^P}{DI_t^P} < 0.84$, then all the terms associated with α_3, α_4, and α_5 will be triggered. Again, as the price ratio falls and these thresholds are crossed, we expect consumer response to become more elastic. In this way, the different levels of discretionary biodiesel use can be captured. We also assume that $\alpha_3 < \alpha_4 < \alpha_5$, which implies that as the price of biodiesel falls relative to the price of diesel, the use of biodiesel increases at an increasing rate, leading to the "kinked" demand curve shown in Fig. 9.1. Total country-specific biodiesel use is further estimated as the product of Q_t^{BD} and total population of that country.

The multimarket, multiregion, dynamic equilibrium models of the international and US agricultural commodity, livestock, and dairy markets are linked to the corresponding biofuels model and solved simultaneously (shown in Fig. 9.2). All these models are based on the principle of structural economic modeling theory as outlined earlier in this section. In this way, we can trace the effects of policy shocks from initial changes in biofuel prices through the resulting changes in feedstock demand and, finally, the changes in food and feed prices triggering changes in the production, consumption, trade, and stocks of the agricultural commodities. The changes in biofuel prices could result in both direct and indirect land use change (LUC). For example, increasing biodiesel mandates in the United States could lead to increases in the use of soybean oil-based biodiesel, which

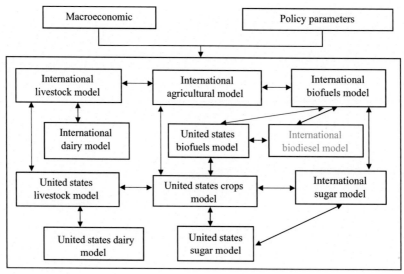

Fig. 9.2 Linkage between international biodiesel model and international agricultural, livestock, dairy, and sugar modeling system.

creates higher demand for soybeans. The higher soybean price, in turn, encourages producers to allocate more land to soybean acreage (direct LUC) and less land toward maize and other competitive crop acreage (indirect LUC) around the world. That could lead to an increase in the global maize price as well. Beyond its use in developing an international biofuels baseline outlook (Debnath et al., 2018), the model can be used to explore several biofuel policy scenarios related to changes in biodiesel policies in the United States and the European Union.

3.1 Baseline domestic biodiesel mandate and trade policy assumptions

The baseline 10-year international biofuel outlook that relies on the key US and EU biodiesel mandate and trade assumptions consistent with current policies is as follows:

1. The US RFS is implemented with requirements corresponding to the rule announced in November 2017 (EPA, 2017). This rule set the biomass-based diesel requirement at 1.75% (2.1 billion gallons) of total transportation fuel use for 2019. We assume it increases linearly to 1.85% (2.3 billion gallons) by 2027, which corresponds to about a 4.7% average biodiesel blend rate.
2. For the European Union, we assume 10% of total transportation fuel is derived from renewable fuels and further assume a 7% cap on food-based biofuels.
3. In terms of biodiesel trade, we assume US antidumping duties on Argentinean biodiesel and no antidumping duties on biodiesel exported to the European Union.

Information regarding further policy assumptions can be obtained from Debnath et al. (2018).

4. Alternative biofuels policy in the United States and the European Union

There are many studies, including Thompson et al. (2010), Broch et al. (2013), Taheripour and Tyner (2013), Tokgoz and Laborde (2014), Keeney and Hertel (2009), and Searchinger et al. (2008), that have discussed the direct and indirect land use consequences of biofuel policies. However, none of these studies explicitly considers the biodiesel policies. In this chapter, we simulate the following biofuel policies that specifically target biodiesel production.

According to the US EPA (2017) and the EU Commission (2015), qualifying biodiesel reduces GHG emissions by at least 50% relative to 2005 base levels of diesel in the United States and 40%–58% in the European Union when vegetable oil-based feedstocks are used. The GHG savings from the use of biodiesel derived from used cooking oil could reach as high as 88%. The United States considers biodiesel derived from vegetable oil as an advanced biofuel in contrast to the European Union, which excludes biodiesel derived from food-based feedstocks from their advanced biofuel lists. Considering this, our main focus in this study is to simulate biodiesel policies that distinguish between the food and nonfood feedstocks in the production of biofuels. Two scenarios are simulated to achieve the aforementioned objectives compared to the baseline:

1. *Scenario US*: The US biodiesel mandate is further increased to 2.88% of transportation fuels by 2027, which corresponds to a volumetric requirement for biomass-based diesel of 3.5 billion gallons compared to 1.85% (i.e., 2.3 billion gallons) by 2027 in the baseline. This corresponds to an average blend rate in diesel fuels of about 6.1% in 2027. We further assume similar growth in the RFS requirement for advanced biofuels, and we assume the portion of the overall mandate that can be met with conventional ethanol remains at 15 billion gallons. We simulate this scenario to test the impact of increased biodiesel demand on the agricultural commodity markets.

2. *Scenario EU*: The current 7% cap on food-based feedstock use in biodiesel production in 2020 is reduced to 5% by 2027 at a rate of 0.3% points per year. The purpose of this policy simulation is to test the impact of decreased food-based feedstock demand on the agricultural commodity market. The scenario assumes that production and use of nonfood-based biodiesel will offset the reduction in food-based biodiesel keeping the overall biodiesel mandate unchanged.

5. Results and discussion

A change in biofuel policy mandates in developed countries can have global consequences. The diversion of food-based feedstocks to biofuel production might raise the price of major food commodities. On the other hand, replacing food-based feedstocks with nonfood-based feedstocks in the production process might lead to a decrease in

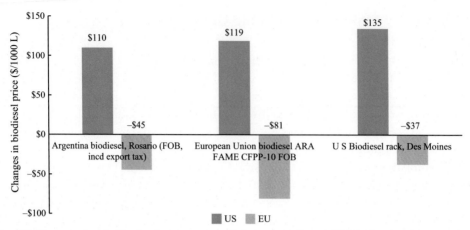

Fig. 9.3 Changes in biodiesel prices corresponding to the US and EU biodiesel policy scenarios. *(Source: Authors' calculations).*

the price of those food crops. In the first scenario, the increase in the US biodiesel mandate from 2.3 billion gallons to 3.5 billion gallons in 2027 resulted in an increase in the US biodiesel rack price by \$135/thousand liters (Fig. 9.3). At those higher prices, the United States also increases biodiesel imports, and the additional demand for Argentinean biodiesel (from the United States or elsewhere) raises Argentinean biodiesel prices by \$110/thousand liters. The global competition for biodiesel also leads to an increase in the EU biodiesel price by \$119/thousand liters.

In the EU scenario, the reduction in demand for food-based feedstocks in biodiesel production results in lower prices for vegetable oil. These reduce the cost of producing vegetable oil-based biodiesel, so the biodiesel price declines relative to the baseline. The US, Argentinean, and EU biodiesel prices fall by \$37/thousand liters, \$45/thousand liters, and \$81/thousand liters, respectively (Fig. 9.3). The relative magnitudes of the US and EU scenarios are, of course, contingent on the particular parameters of the alternative scenarios considered.

Higher biodiesel demand in the United States raises the wholesale price of biodiesel to spur additional production, and the increased demand for soybean oil leads to higher global prices across the vegetable oil complex (Table 9.1). Most biodiesel produced in the United States is derived from soybean oil. Increasing the US biodiesel mandate in the first scenario leads to an increase in the production of biodiesel by 2463 million liters, which increases the use of soybean oil for biodiesel production in the United States by 1462 thousand metric tons (Table 9.2). Biodiesel production in Argentina and the European Union decreases as the relative ratio of output (biodiesel) prices to input (vegetable oil) prices falls, leading to lower utilization of the production capacity in those countries (Tables 9.1 and 9.2).

Table 9.1 Changes in grains and oilseed (meal and oil) prices under US and EU biodiesel policy scenarios (3-year average of 2025–27)

Changes in grains, oilseed crops, oil, and meal prices ($/metric ton)		
	Scenarios	
	US	EU
Grains		
Maize	1.25	0.24
Wheat	1.60	−1.34
Rice	0.60	−0.29
Sorghum	0.85	−0.03
Oilseed, oil, and meal		
Soybeans	1.99	1.97
Oil	107.04	−40.06
Meal	−23.51	13.29
Rapeseed	15.83	−24.21
Oil	81.11	−102.07
Meal	−31.18	27.44
Sunflower	10.94	−7.25
Oil	67.93	−42.98
Meal	−33.87	19.95

Note: US Gulf, yellow No. 2 maize price; Rotterdam, No. 1 dark northern spring wheat price; Decatur Illinois, soybean meal and oil prices; FOB Thailand, 100% Grade B rice price; US Gulf, No. 2 sorghum price; Central Illinois, No. 1 yellow soybean price; Hamburg, rapeseed and meal prices; Rotterdam, rapeseed oil price; Rotterdam, sunflower seed and meal prices; NW Europe, sunflower seed oil price.
Source: Authors' calculations.

In the baseline, a small portion of the biodiesel and other advanced biofuels are not only required to meet their respective mandates, but also used to help satisfy the overall RFS. This happens because it is easier to increase biodiesel use than it is to encourage the use of additional maize-based ethanol in blends higher than the conventional 10% blend with gasoline. In the scenario, the increase in the biodiesel and advanced mandates is large enough that it effectively squeezes out this use of biodiesel to satisfy the broader biofuel use mandate. This means more of the overall requirement must be met with other fuels, such as conventional ethanol, so ethanol production rises in the United States and Brazil as well. It may seem odd that an increase in a biodiesel mandate has the practical effect of also increasing demand for ethanol, but this illustrates the complex market interactions that result from the RFS and that are captured by the model.

In the EU scenario, on the other hand, less use of food-based feedstocks leads to falling vegetable oil prices. Cheaper vegetable oil prices encourage Argentina and the United States to produce more biodiesel even though the biodiesel price decreases (Fig. 9.3 and Table 9.1). The findings in this scenario are somewhat surprising at first glance, but they match our expectations of the cross-effects that occur in a global market.

Table 9.2 Changes in biodiesel and ethanol production and feedstock use for major biofuels-producing countries; US and EU biodiesel policy scenarios (3-year average of 2025–27)

| Countries | Production (million liters) | | Feedstock use by type (1000 metric tons) | | |
| | Scenarios | | | Scenarios | |
	US	EU	Feedstock type	US	EU
US					
Biodiesel	2463	107	Soybean oil	1462	−16
			Maize oil	207	−9
			Other oil	895	181
Ethanol	880	−178	Maize	3206	−658
Brazil					
Ethanol	165	−24	Sugarcane	2064	−295
Argentina					
Biodiesel	−316	47	Soybean oil	−321	48
EU					
Biodiesel	−48	278	Rapeseed oil	−24	−1204
			Other nonfood	−26	1481
Indonesia					
Biodiesel	325	−53	Palm oil	298	−48

Source: Authors' calculations.

By reducing the share of biodiesel use in the European Union that can be derived from food-based vegetable oils, such as rapeseed oil, the prices for those feedstocks fall (Table 9.1). Lower feedstock prices moderate the drop in EU production of biodiesel from rapeseed oil and other vegetable oils, and it is assumed that the supply of nonfood-based feedstocks is sufficiently elastic that the increase in production of biodiesel from these feedstocks can more than offset the reduction in production from vegetable oil. With total biodiesel production increased and use essentially determined by the policy, the result is reduced biodiesel imports into the European Union (Table 9.3). Biodiesel production in the United States increases (Table 9.2) in response to lower feedstock prices and higher net returns that more than compensate for the lower biodiesel price. Because there is additional biodiesel available in the United States to meet the broader requirement, there is a slight substitution away from ethanol to meet the broader requirements and less ethanol production as a result (Table 9.1).

In Table 9.1, we further trace the impact of changes in biodiesel policies on vegetable oil, meal, oilseed crop, and grain prices. We find that the increase in the US biodiesel mandate leads to an increase in biodiesel feedstock prices around the globe relative to

Table 9.3 Changes in biodiesel trade in the case of the US and EU scenario

Countries	Net trade (million liters)	
	Scenarios	
	US	EU
Net import		
US	56	41
EU	48	−281
Net export		
Argentina	−307	36
Indonesia	323	−53

Source: Authors' calculations.

the baseline. The world soybean oil price increases by $107/metric ton, the rapeseed oil price increases by $81/metric ton, and the sunflower oil price increases by $68/metric ton. Similar impacts can be traced to the grain markets too, where grain prices rise with a higher US biodiesel mandate. This occurs because substitution toward additional ethanol production leads to higher maize prices. The increases in vegetable oil and grain prices have food security implications, increasing the costs of staple foods and increasing live-stock feed costs.

In contrast, oilseed-based meal prices decline in the US policy scenario. The increase in oilseed production that results from higher prices for vegetable oils and oilseeds increases the supply of meal. The resulting lower prices for soybean meal and other oilseed-based high-protein meals more than offsets the increase in livestock feed costs that results from higher grain prices. The effect would be especially pronounced for poul-try rations that utilize a high proportion of oilseed meal, but lower meal prices have little effect on cattle feed ration costs. Changes in feed costs are ultimately reflected in the prices of meat and dairy products. Thus a policy that increases demand for biodiesel can have the effect of raising consumer prices for vegetable oils and the many products made with vegetable oils, but reducing prices for poultry and other animal-based foods.

When the model is simulated with the assumption of less use of food-based feedstocks in biofuel production in the European Union, we find vegetable oil prices, including soybean oil, rapeseed oil, and sunflower oil, drop by $40/metric ton, $103/metric ton, and $43/metric ton, respectively, while grain market prices respond differently with a minor increase in maize prices (Table 9.1). This occurs because the relative returns to crop production favor additional oilseed area (primarily for soybeans) at the expense of maize and other grains. The impact on oil prices is highest for rapeseed oil because the majority of European Union's biodiesel is produced from rapeseed oil and the limit on food-based feedstock use falls on rapeseed oil the hardest. With the reduction in vegetable oil prices, fewer oilseeds are crushed and the availability of meal to the livestock sector falls, which causes higher meal prices (Table 9.1). Responding to this downward pressure in the vegetable oil market, both world rapeseed and sunflower seed prices fall by $24/metric ton and $7/metric ton, respectively. However, soybeans have a lower oil content than rapeseed and sunflowers, so in this case, the increase in the value of the meal component offsets the decrease in the value of the oil. As a result of the better crush margin, soybean prices are slightly higher by $2/metric ton, and that leads to some substitution of soybean area for maize area.

6. Summary and conclusion

The analysis and policy discussions in the context of biodiesel mandates are complex because those policies are directly linked to vegetable oil markets and can indirectly impact grain markets. We shed some light on how changes in US and EU biodiesel policy can shift the agricultural commodity prices that have food security consequences. An increase in the US biodiesel mandate could lead to upward pressure on vegetable oil and grain prices. While restricting the use of food-based feedstocks in biodiesel used in the European Union would have the opposite effect. Both scenarios therefore have consequences for global food security. In both cases, however, the consumer-level impacts of changes in vegetable oil and grain prices are at least partially offset by oilseed meal prices that move in the opposite direction.

Of course, the scenarios also have implications for food producers. Higher oilseed and grain prices would increase farm income in the US policy scenario, while the EU scenario would reduce producer margins. While the impacts on farm income in these scenarios are relatively modest, they would increase the vulnerability of low-income farmers that produce these products. Although not explored here, the increase or decrease in demand for biodiesel feedstocks, and the resulting changes in those feedstock prices, could lead to both direct and indirect LUC. For these scenarios, the changes in global area devoted to grains and oilseeds are modest because the changes in grain and oilseed prices are small.

Modeling the nonfood biodiesel feedstock (used cooking oil, animal fat, etc.) sector is beyond the scope of this study. We assume a perfectly elastic supply of nonfood-based feedstocks. However, this may not be the case in reality. The significant increase in the use of nonfood-based feedstock in the EU scenario would limit other existing uses for those products, which could have implications for food markets.

A lesson from the study is that the implications of changes in biodiesel policy are complex. To estimate impacts, models must consider the nuances of policies and the interactions of multiple markets. The effects of expanding biodiesel use may be similar to the effects of increasing ethanol consumption in many respects, but there are also many important differences as highlighted in this study.

References

Abbott, P., Hurt, C., Tyner, W., 2008. What's Driving Food Prices? Farm Foundation.

Babcock, B., 2011. The impact of US biofuel policies on agricultural price levels and volatility. In: Prepared for the International Centre for Trade and Sustainable Development, Geneva, Switzerland.

Barr, K., Babcock, B., Carriquiry, M., Nassar, A., Harfuch, L., 2011. Agricultural land elasticity in the United States and Brazil. Appl. Econ. Perspect. Policy 33 (3), 449–462.

Broch, A., Hoekman, S.K., Unnach, S., 2013. A review of variability in indirect land use change assessment and modeling in biofuel policy. Environ. Sci. Pol. 29, 147–157.

Chakravorty, U., Hubert, M., Nøstbakken, L., 2009. Fuel versus food. Annu. Rev. Res. Econ. 1 (1), 645–663.

Chen, X., Huang, H., Khanna, M., Onal, H., 2011. Meeting the Mandate for Biofuels: Implications for Land Use, Food and Fuel Prices. (NBER Working Paper No. 16697).

Debnath, D., Whistance, J., Thompson, W., 2017a. The causes of two-way US-Brazil ethanol trade and the consequences for greenhouse gas emission. Energy 141, 2045–2053.

Debnath, D., Whistance, J., Thompson, W., Binfield, J., 2017b. Complement or substitute: ethanol's uncertain relationship with gasoline under alternative petroleum price and policy scenarios. Appl. Energy 191, 385–397.

Debnath, D., Westhoff, P., Whistance, J., Thompson, W., Binfield, J., 2018. International Biofuels Baseline Briefing Book. FAPRI-MU Report 2-18. University of Missouri, Columbia.

Dumortier, J., Hayes, D.J., Carriquiry, M., Dong, F., Du, X., Elobeid, A., Fabiosa, J.F., Tokgoz, S., 2011. Sensitivity of carbon emission estimates from indirect land-use change. Appl. Econ. Perspect. Policy 33 (3), 428–448.

Egbendewe-Mondzozo, A., Swinton, S.M., Kung, S., Post, M.W., Binfield, J.C., Thompson, W., 2015. Bioenergy supply and environmental impacts on cropland: insights from multi-market forecasts in a Great Lakes subregional bioeconomic model. Appl. Econ. Perspect. Policy 37 (4), 602–618.

EIA (Energy Information Administration), 2018. Monthly Energy Review Appendix A: British Thermal Unit Conversion Factor.

EISA, 2007. Energy Independence and Security Act of 2007. Public Law 110–140, 121 Stat. 1492, HR6, United States. www.gpo.gov/fdsys/pkg/BILLS-110hr6enr/pdf/BILLS-110hr6enr.pdf.

EPA (Environmental Protection Agency), 2017. Renewable Fuel Standard Program: Standards for 2018 and Biomass-Based Diesel Volume for 2019.

EU Commission, 2015. Directive 2009/28/EC of the European parliament and of the council. Off. J. Eur. Union (2009L0028—EN—05.10.2015—002.001—1).

Ewing, M., Msangi, S., 2009. Biofuels production in developing countries: assessing tradeoff in welfare and food security. Environ. Sci. Pol. 12, 520–528.

F.O. Licht Database, 2018. World Ethanol and Biodiesel Production, Use, Trade, and Stocks Data.

FAO, 2012. World Agriculture Towards 2030/2050: The 2012 Revision. ESA Working Paper No. 12-03.

FAO (Food and Agricultural Organization of the United Nations), 2008. The State of Food and Agriculture e Biofuels: Prospects, Risks and Opportunities. The Food and Agriculture Organization of the United Nations, Rome, Italy.

Hochman, G., Rajagopal, D., Zilberman, D., 2010. Are biofuels the culprit: OPEC, food, and fuel. Am. Econ. Rev. (2), 183–187.

Hochman, G., Rajagopal, D., Timilsina, G., Zilberman, D., 2014. Quantifying the causes of the global food commodity price crisis. Biomass Bioenergy 68, 106–114.

Keeney, R., Hertel, T.W., 2009. The indirect land use impacts of United States biofuel policies: the importance of acreage, yield, and bilateral trade responses. Am. J. Agric. Econ. 91 (4), 895–909.

Meyer, S., Thompson, W., 2012. How do biofuel use mandates cause uncertainty? United States Environmental Protection Agency cellulosic waiver options. Appl. Econ. Perspect. Policy 34 (4), 570–586.

Müller, A., Schmidhuber, J., Hoogeveen, J., Pasquale, S., 2008. Some insights in the effect of growing bioenergy demand on global food security and natural resources. Water Policy 10 (S1), 83–94.

Naylor, R., Higgins, M., 2018. The rise in global biodiesel production: implications for food security. Glob. Food. Secur. 16, 75–84.

OECD-FAO, 2017. OECD-FAO Agricultural Outlook 2017–2026. OECD and FAO Secretariats, Organization for Economic Cooperation and Development/Food and Agriculture Organization United Nations, Rome, Italy.

Pingali, P., Raney, T., Wiebe, T., 2008. Biofuels and food security: missing the point. Appl. Econ. Perspect. Policy 30 (3), 506–516.

Rosegrant, M., 2008. Biofuels and Grain Prices: Impacts and Policy Response. Testimony for the US Senate Committee on Homeland Security and Governmental Affairs.

Searchinger, T., Heimlich, R., Houghton, R.A., Dong, F., Elobeid, A., Fabiosa, J., Yu, T.-H., et al., 2008. Use of US croplands for biofuels increases greenhouse gases through emissions from land-use change. Science 319 (5867), 1238–1240.

Serra, T., Zilberman, D., JM, G., 2011a. Price volatility in ethanol markets. Euro Rev. Agric. Econ. 38 (2), 259–280.

Serra, T., Zilberman, D., Gil, J.M., Goodwin, B.K., 2011b. Nonlinearities in the US corn-ethanol-oil Price system. Agric. Econ. 42, 35–45.

Taheripour, F., Tyner, W.E., 2013. Biofuels and land use change: applying recent evidence to model estimates. Appl. Sci. 3, 14–38.

Thompson, W., Meyer, S., 2013. Second generation biofuels and food crops: co-products or competitors? Glob. Food. Secur. 2 (2), 89–96.

Thompson, W., Meyer, S., Westhoff, P., 2010. The new markets for renewable identification numbers. Appl. Econ. Perspect. Policy 32 (4), 588–603.

Thompson, W., Whistance, J., Meyer, S., 2011. Effects of US biofuel policies on U.S. and world petroleum product markets with consequences for greenhouse gas emission. Energy Policy 39, 5509–5518.

Timilsina, G.R., Shrestha, A., 2011. How much hope should we have for biofuels? Energy 36 (4), 2055–2069.

Tokgoz, S., Laborde, D., 2014. Indirect land use change debate: what did we learn? Curr. Sustain. Renew. Energ. Rep. 1, 104–110.

USDA-FAS (US Department of Agriculture-Foreign Agricultural Service), 2015a. India Biofuels Annual. GAIN Report # IN5079.

USDA-FAS (US Department of Agriculture-Foreign Agricultural Service), 2015b. China Biofuels Annual. GAIN Report # CH15030.

USDA-OCE (US Department of Agriculture-Office of the Chief Economists), 2018. Long-Term Agricultural Projections. https://www.usda.gov/oce/commodity/projections/. (Accessed 30 March 2018).

Whistance, J., Thompson, W., 2014. A critical assessment of RIN Price behavior and the implications for corn, ethanol, and gasoline Price relationships. Appl. Econ. Perspect. Policy 36 (4), 623–642.

Wright, B., 2014. Global biofuels: key to the puzzle of grain market behavior. J. Econ. Perspect. 28 (1), 73–98.

Zhang, Z., Lohr, L., Escalante, C., Wetzstein, M., 2010. Food versus fuel: what do prices tell us? Energy Policy 38, 445–451.

Further reading

Collins, K., 2008. The Role of Biofuels and Other Factors in Increasing Farm and Food Prices: A Review of Recent Developments With a Focus on Feed Grain Markets and Market Prospects. University of California, Davis.

Food and Agricultural Policy Research Institute, 2018. US Baseline Outlook. FAPRI-MU Report #01–18.

Mitchell, D., 2008. A Note on Rising Food Prices. The World Bank Development Prospects Group, Washington, DC.

CHAPTER 10

The impact of key US biofuel policies: An example of the US RFS and California's LCFS

Jarrett Whistance, Wyatt Thompson
Food and Agricultural Policy Research Institute, University of Missouri, Columbia, MO, United States

Contents

1. Introduction

Currently, there are two primary renewable fuel polices at work in the United States. At the federal level is the Renewable Fuel Standard (RFS) that requires specific volumes of renewable fuel use each year based in part on the level of greenhouse gas (GHG) emission reductions provided by broad categories of fuel. At the state level, California's Low Carbon Fuel Standard (LCFS) has a similar aim to reduce GHG emissions but achieves that goal through a different design. The LCFS lays out a path of carbon intensity (CI) reductions in fuel use but does not prescribe specific volumes of fuel to meet those reductions. Both policies have similar enforcement mechanisms: Renewable Identification Numbers (RINs) for the RFS and CI credits for the LCFS are tradable certificates that obligated parties use to demonstrate compliance with the policy. The relative degrees to which the policies are binding can make for some interesting interactions. Those interactions are what we explore in this chapter.

The RFS was first established as part of the Energy Policy Act of 2005 and was later revised and expanded as part of the Energy Independence and Security Act of 2007. In its expanded form, the RFS became a hierarchical mandate with a broad requirement for renewable fuels that met at least a 20% GHG emission reduction relative to petroleum-based fuels and submandates for more advanced biofuels that met at least a 50% GHG reduction threshold, including biomass-based diesel and cellulosic fuels, which required at least a 60% reduction. While the original requirements called for

Biofuels, Bioenergy and Food Security
https://doi.org/10.1016/B978-0-12-803954-0.00010-3

36 billion gallons of renewable fuel to be used by 2022, of which 16 billion gallons were supposed to be sourced from cellulosic feedstocks and 15 billion gallons at most could be sourced from conventional feedstocks like corn starch, the requirements to date have been reduced substantially to account for the slower than expected development of cost-competitive cellulosic technology. For example, the requirements for 2018 were set at a little over 19 billion gallons, of which the cellulosic requirement comprised only a few hundred million gallons (EPA, 2017).

Each year, the US Environmental Protection Agency sets the requirements for the following year. The standards are finalized as the percent of estimated motor fuel use that corresponds to a targeted volume of renewable fuel use. Obligated parties (e.g., domestic oil refineries and motor fuel importers) apply the percent standards to their output during the compliance year and submit the requisite number of RINs to demonstrate compliance. RINs are generated alongside renewable fuel as it is produced or imported. They remain associated with a particular quantity of renewable fuel until it is blended for inclusion into the domestic motor fuel supply when they become "separated." Obligated parties can obtain RINs for compliance by physically blending the appropriate amount of renewable fuels into their output for domestic use or by purchasing surplus separated RINs from other obligated parties or RIN holders. Excess RINs, up to 20% of the following year's requirement, may also be used to meet the requirements in the next compliance period.

The LCFS was established in California in 2010 and went into effect the following year. The original goal of the legislation was to reduce by 10% the CI of motor fuels used in California by 2020. CI reductions were small in the early years but have been on an accelerated pace at the time of writing to meet the 10% goal. The LCFS differs from the RFS in that it does not specify certain amounts of renewable fuel that must be used. Instead, the LCFS assigns CI scores to individual fuels based on their production process. Obligated parties must supply fuel with an average CI equal to the requirement spelled out by legislation. If the average CI of the fuel they supply is more carbon intense, they generate deficits that must be met with additional LCFS credits. If the fuel provided is less carbon intense, those parties generate surplus credits that may be sold. This flexibility gives obligated parties the option to blend whichever optimal mix of fuels satisfies their CI reduction obligations or they may purchase LCFS credits from other obligated parties. Excess CI credits may be carried forward for up to 5 years.

In both cases, complying with the policies comes at a cost. The magnitude of this cost depends on the degree to which the policies are binding. In the case of the RFS, a binding mandate occurs when the annual volume obligation exceeds the amount that would have been chosen by the market in the absence of the policy (Fig. 10. 1). This creates a price wedge, which we refer to as the "core RIN value," between the price blenders must pay to induce biofuel producers to supply fuel at the wholesale level and the price at which

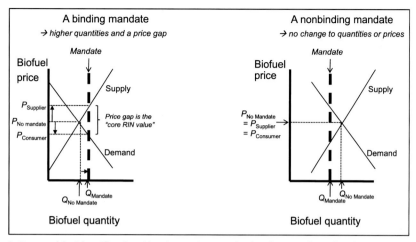

Fig. 10. 1 Renewable Identification Number prices under binding and nonbinding Renewable Fuel Standard requirements. *(From Thompson, W., Meyer, S., Westhoff, P., 2009. Renewable identification numbers (RINs) are the bellwether of US biofuel mandates. Eurochoices).*

they must sell to induce consumers to use the fuel at the retail level. The more binding the mandate, the larger the price wedge and the higher the RIN value.

In early years when the RFS was not binding, conventional RINs could be purchased for only a few pennies each. As the requirements grew and reached levels that could not be attained with 10% blends of ethanol and motor gasoline alone, RINs became much more expensive. They reached a peak of $1.46 per RIN in 2013. RIN prices remain quite volatile, but they have not settled to the lows seen prior to 2013 at least through the middle of 2018. LCFS credits, likewise, have seen increases in price during the first several years of the existence of this policy. As the CI requirements become more binding, credits are harder to come by and their prices rise. At the time of writing, LCFS credits are in record territory of more than $180 per credit.

These increases in price have led to renewed calls to look at the policies in greater detail and perhaps to study potential reforms. Opponents of these policies argue that the additional costs pose a burden to obligated parties. Supporters, on the other hand, posit that the higher prices are evidence that the policies are working as intended (i.e., higher RIN and LCFS credit prices encourage additional renewable fuel use in the form of higher blends, like E85, or through the use of "greener" fuels that generate more credits).

By themselves, each policy has garnered quite a bit of attention in the academic literature, with most studies focusing the individual structure of the policies and perhaps their effects on related agricultural and energy markets or on their ability to meet certain climate or GHG reduction goals. Fewer studies have investigated the interaction between the two policies, and at least some of those that have done so use a hypothetical federal-level LCFS representation instead of a state-level representation (Huang et al., 2013). By and large, the results published to date indicate a degree of interdependence between the policies, such that the policies tend to have a greater impact when working in tandem than either one would have in isolation (Christensen and Hobbs, 2016; Rubin and Leiby, 2013; Whistance et al., 2017).

One concern with the state-level representation is that the policy structure might lead to a situation of fuel shuffling, in which the more environmentally-friendly fuels are used in one area but have not displaced the less environmentally-friendly fuels from being used elsewhere. Thus the policies might not lead to an actual decline in GHG emissions; the emissions are just "shuffled" elsewhere. Some studies have noted this possibility because it relates to the RFS and ethanol trade with Brazil (Rajagopal, 2015). There is also the potential that intraregional biofuel trade can affect estimates of environmental effects of biofuels, including emissions (Yang, 2018).

It is clear that each of these policies affects motor fuel and other agricultural commodity markets to varying degrees. Although the effects of each of these policies have been studied in detail in the academic literature, the need for additional research is evidenced by the vigor of these debates. The focus of this chapter is to explore how the policies and their effects interact with one another. This will give readers a better understanding of the role of policy design in market outcomes.

2. Method

To achieve our objective, we modify an existing partial-equilibrium model of US and international agriculture and energy markets (for model details, see Whistance and Thompson, 2014; Whistance et al., 2017). The Food and Agricultural Policy Research Institute at the University of Missouri (FAPRI-MU) modeling system is tailored for policy analysis and is updated twice a year, generally, to generate 10-year projections of market outcomes under current policy assumptions. These projections serve as a benchmark, or baseline, from which additional policy analysis can proceed. By changing policy assumptions and estimating the changes in market prices and quantities that occur relative to the baseline, analysts can provide information to policymakers regarding the magnitude and direction of potential policy impacts.

Because the focus of this chapter is on measuring the impact of policies at two different spatial levels (e.g., national and subnational), the biofuel model is modified to model the US biofuel markets at a subnational level that can then be aggregated to the broader national level. This allows for a more explicit representation of the LCFS policy in California, in addition to the RFS representation that was already a part of the model.

The primary modifications take place on the demand side, because the LCFS and RFS are demand oriented. Motor fuel demand is separated into two regions: California and the Rest-of-US (ROUS). Prices are allowed to differ by region as the policies and market conditions warrant. The share of renewable fuels is also allowed to differ because a binding LCFS might lead to a greater market share of high-blend fuels like E85 and B20 in California versus the ROUS.

The renewable fuels themselves are modeled somewhat differently between the regions. Whereas the ROUS region relies on the broad RFS hierarchy of conventional, advanced, and cellulosic fuels, the California region uses a somewhat more refined LCFS hierarchy that splits both ethanol and biomass-based diesel into four subcategories based on different CI ranges. The CI ranges are determined based on the average values listed by the California Air Resources Board and not based on the individual pathway approvals (CARB, 2015). CI ratings are assumed to be price responsive, so higher LCFS credit prices motivate further improvements in production processes to capture more of that value. However, the model does not allow categories to "leapfrog" one another. For example, the CI rating for conventional ethanol is allowed to improve, but it cannot attain a better CI rating than cellulosic ethanol. This assumption is necessary to prevent instability in the model solution process and, in most cases, is a technical step that is not required given the relative CI ratings.

Another simplifying assumption that is made on the demand side relates to the order in which the fuels are used. In this case, we assume fuels in California are used in the order of their CI ranking. Thus fuels that are less carbon intense (and generate greater LCFS credits) will be used before fuels that are more carbon intense (and generate fewer LCFS credits). The implication is that some fuels might be shipped to and consumed almost entirely in California with little to no use of those fuels occurring in the ROUS region. Anecdotally, this seems to be the case with ethanol imported to the United States from Brazil for use almost exclusively in California, but it may not be the case for all renewable fuels produced in the United States. Here we assume that once 75% of a particular fuel (e.g., cellulosic ethanol) is used in California, the transition is made to the fuel in the next best CI range. In the case of ethanol, for example, the transition would occur from cellulosic ethanol to advanced ethanol.

Similar to the representation of the RFS and RIN prices in the FAPRI-MU model, the additional model for LCFS credits is structured so that the LCFS credit price is determined at the point where the supply of and demand for credits balance. The RFS and LCFS credit prices both create wedges that lead to cross-subsidization of less carbon intense renewable fuels at the expense of more carbon intense fuels. The cross-subsidization also encourages higher market shares for fuels with greater concentrations of renewable fuel (e.g., E85 and B20).

Another key difference on the demand side relates to the stockholding behavior of obligated parties with respect to RINs and LCFS credits. Both policies allow obligated parties to carry credit stocks forward for a limited length of time. However, the lengths of time differ by policy. Under the RFS, RINs may be carried forward one additional year to meet up to 20% of the following year's obligation. Thus RINs produced in a given year must be used in the same year or the following year, or else they expire. LCFS credits, on the other hand, can be carried forward for up to five years before they expire. The longer "shelf life" of LCFS credits is an important consideration, because stocks can be built if CI requirements are easier and drawn down over time. This might lead to somewhat more stable LCFS credit prices over time.

On the supply side, the structure of the aggregate renewable fuel production equations is unchanged from previous versions of the FAPRI-MU model. Renewable fuel production is a function of the wholesale prices and marginal net returns. The wholesale prices capture both the marginal value of that particular fuel in meeting the RFS (based on the endogenous RIN prices) and the weighted marginal value of that fuel meeting the LCFS in California (based on the endogenous LCFS credit prices and the share of that fuel consumed in California).

To analyze the interaction between the RFS and the LCFS, we run four pairs of scenarios. Three pairs correspond to the three assumed paths for RFS requirements: No RFS; Low RFS (requirements grow by 1.9% per year); and High RFS (requirements grow by 2.6% per year), and the fourth pair adds a higher assumed crude oil price path on top of the High RFS case. Within each pair, we compare two cases: a No LCFS scenario and a Baseline LCFS scenario. This gives us a spectrum of policy assumptions that range from a "No Policy" scenario (No RFS; No LCFS) to a "Current Policy" scenario (Low RFS; Base LCFS) and finally to a "Stringent Policy" scenario (High RFS; Base LCFS). Fig. 10. 2 shows the path of these alternative policy assumptions over time, and Fig. 10. 3 shows the alternative crude oil price paths.

3. Results

In this section, we lay out the results starting with motor fuel use effects and moving toward agricultural market effects and policy compliance costs. Figs. 10. 4 and 10.5 illustrate the motor fuel use effects in aggregate. It is not readily apparent from the figure, but the imposition of renewable fuel policies leads to a decrease in overall motor gasoline use. Motor gasoline use is highest in the "No Policy" case and decreases with the introduction of LCFS requirements (within-pair comparison) and with the introduction of the RFS (across-pair comparison). More noticeable is that the high crude oil price assumption leads to a very steep decline in motor gasoline use over time. Within each pair, the introduction of the LCFS reduces motor gasoline use in California and that reduction appears in the aggregate measure as well. There is not much of an offsetting increase in ROUS motor gasoline use. Across the pairs, motor gasoline use declines with the introduction of the "Low RFS" requirements but does not decrease further with a "High RFS" assumption unless we assume a higher crude oil price path.

Renewable fuel use increases, in aggregate, as the scenarios move from "No Policy" to "High RFS; Base LCFS." Across scenarios, we see the ethanol and biodiesel uses increase from 13.5 billion gallons and 1.5 billion gallons to 17.5 billion gallons and 3.2 billion gallons, respectively (Fig. 10. 5). Although there is a 15 billion gallon cap

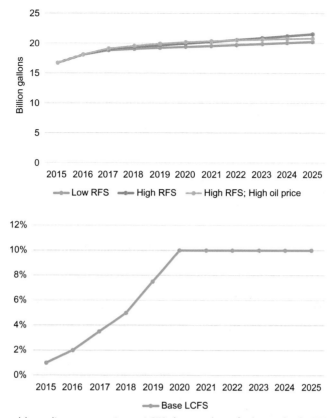

Fig. 10. 2 Renewable policy assumptions. *LCFS*, low carbon fuel standard; *RFS*, renewable fuel standard. *(Source: Authors' calculation).*

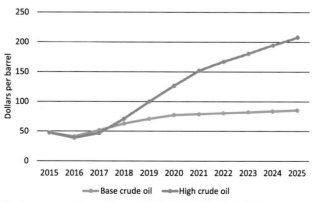

Fig. 10. 3 Crude oil price assumptions. *(From IHS Markit, 2016. Global Macroeconomic Outlook; Energy Information Administration, 2018. Annual Energy Outlook 2018: High Oil Price Case).*

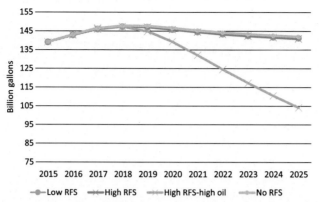

Fig. 10. 4 Motor gasoline use, 2015–25. *Note*: the scale on the vertical axis has been changed to highlight the difference between scenarios. *RFS*, renewable fuel standard. *(Source: Authors' calculations)*.

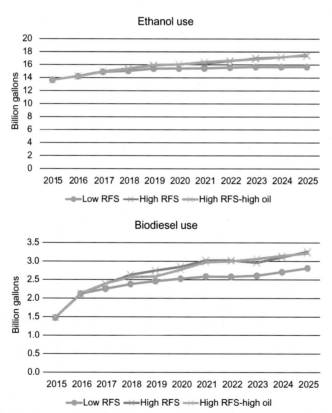

Fig. 10. 5 Renewable fuel use, 2015–25. *RFS*, renewable fuel standard. *(Source: Authors' calculations)*.

on the contribution of conventional ethanol toward the RFS requirements, we see more than 15 billion gallons of ethanol use in the High RFS cases because market conditions are favorable for further advanced biofuel requirements, requiring more sugarcane ethanol from Brazil and cellulosic ethanol production, and further growth in high-level blends like E85. Within scenarios, we see offsetting changes as the LCFS tends to shift ethanol and biodiesel consumption to California and away from the ROUS. These shifts are also apparent in terms of ethanol blend shares.

Ethanol blends, in aggregate, tend to shift away from low-level blends like E10 and toward high-level blends like E85 if RFS and the LCFS are in place and become more stringent (Fig. 10. 6). We see this occur in both California and the ROUS region as well. The offsetting changes in motor gasoline use appear to limit the growth in higher blends within scenarios. For example, the introduction of the LCFS in the "High RFS" pairing

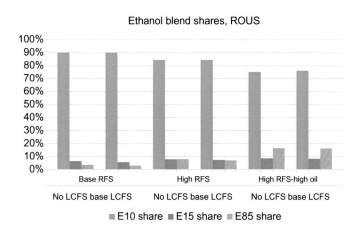

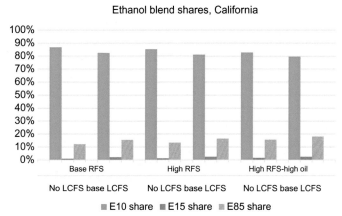

Fig. 10. 6 Ethanol blend market shares, by region, 2015–25 average. *LCFS*, low carbon fuel standard; *RFS*, renewable fuel standard; *ROUS*, rest of US. *(Source: Authors' calculations).*

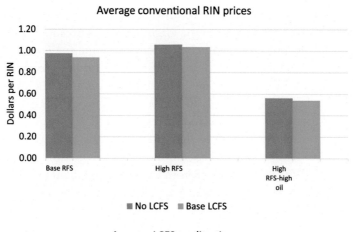

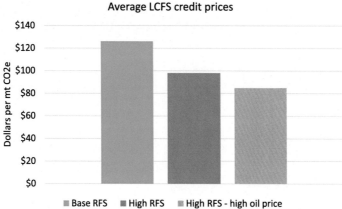

Fig. 10. 7 Compliance credit prices, 2015–25 average. *LCFS*, low carbon fuel standard; *RFS*, renewable fuel standard. *(Source: Authors' calculations).*

leads to increases in the California market share of both E15 and E85 at the expense of E10. At the same time, there is little change in the ROUS market shares for E15 and only slight changes for market shares of E85 within the scenarios.

In Fig. 10. 7, we see how the compliance costs compare across scenarios. It stands to reason that compliance costs for each policy will increase with the introduction or increase in the stringency of that policy. For example, within each pairing the introduction of the LCFS requirements leads to the establishment of LCFS credit prices, and across the pairings we see rising RIN prices as well. The interaction between these policies is evident. Within the low and high RFS pairings, we see that the introduction of the LCFS leads to a corresponding decrease in the RFS compliance costs, as measured by both the individual RIN prices and the overall compliance cost. The addition of high oil prices leads to a reduction in both measures of compliance cost, though the reduction appears greater for RIN prices compared to LCFS credit prices. When looking across scenarios, the LCFS compliance

Table 10. 1 Retail gasoline and implied retail ethanol prices by region, 2015–25 average

Retail gasoline prices ($/gal)	No RFS		Low RFS		High RFS		High RFS; high oil	
	No LCFS	Base LCFS	No LCFS	Base LCFS	No LCFS	Base LCFS	No LCFS	Base LCFS
Rest of US	$2.68	$2.68	$2.78	$2.78	$2.81	$2.80	$3.93	$3.93
California	$3.14	$3.32	$3.24	$3.38	$3.26	$3.36	$4.39	$4.47
Implied retail ethanol price ($/gal)								
Rest of US	$2.39	$2.39	$1.78	$1.80	$1.73	$1.75	$2.26	$2.27
California	$2.39	$2.20	$1.78	$1.63	$1.73	$1.58	$2.26	$2.14

LCFS, low carbon fuel standard; *RFS*, renewable fuel standard.
Source: Authors' calculations.

costs decrease as the RFS requirements become more stringent. The implication is that both policies appear to be mutually reinforcing or complementary. In other words, as one policy becomes more stringent, the efforts of obligated parties to meet that new requirement make the other requirement marginally less cumbersome.

Table 10. 1 shows the effect of these policies on motor fuel prices. There we see the sort of cross-subsidization that we would expect between retail gasoline and ethanol prices. To spur additional renewable fuel use (e.g., ethanol use), both policies act to reduce the implied retail ethanol price and increase the retail gasoline prices. The idea is that the compliance mechanisms act as wedges between the wholesale and retail renewable fuel prices and obligated parties pass the cost of those wedges on to final consumers of the petroleum products. Within each pairing, we see higher retail gasoline costs and lower implied retail ethanol costs in California as the LCFS is introduced. It is interesting to note that the LCFS introduction also leads to somewhat higher ethanol prices in the ROUS. The incentive to pull more ethanol into California for use to meet the LCFS requirement bids up the price for ethanol in other states.

As the RFS requirements are introduced, we see higher retail gasoline costs and lower implied retail ethanol costs in both California and the ROUS. This makes sense because the RFS is a federal policy, whereas the LCFS is just a state-level policy. The complementary nature of the policies is also evident in the motor gasoline price effects. Here we can see within the RFS scenarios that the implementation of the LCFS takes some of the burden off the RFS, so the implied retail ethanol prices in the ROUS increase relative to the "No LCFS" case.

In terms of relative costs, the high oil price assumption can make quite a difference. Fig. 10. 8 illustrates this point. In a High RFS and Base LCFS scenario, moving from the baseline crude oil price assumption to the high crude oil price assumption leads to an increase in relative motor gasoline expenditures by 30% and a decrease in relative policy

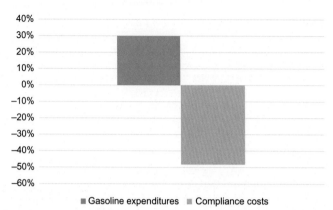

Fig. 10. 8 Change in fuel expenditures and compliance costs, high oil case relative to base, 2015–25 average. *(Source: Authors' calculations).*

compliance costs of nearly 50%. On the one hand, high oil prices mean a steep reduction in motor fuel demand, though this reduction is offset by the higher fuel prices and expenditures rise in total. On the other hand, biofuels are more attractive compared to petroleum fuel products, so both the RFS and LCFS are much easier to meet and compliance costs fall as a result. These estimates assume that use of ethanol in high-blend fuels would expand more rapidly than in the past given these retail prices. Finally, the absolute increase in motor fuel expenditures is much larger than the decrease in compliance costs.

4. Conclusions

We have investigated two of the primary biofuel policies in the United States. On the one hand is a subnational policy, namely the LCFS in California. On the other hand is the RFS at the federal level. Both policies have the explicit goal of reducing GHG emissions, but they use different methods to achieve that goal. The LCFS specifies a path of CI reductions but gives obligated parties the flexibility to determine which mix of fuels is optimal for them to meet their obligations. The RFS, however, spells out the volume and makeup of the renewable fuel mix to which obligated parties must adhere to demonstrate compliance. In this chapter, our aim was to gain a better understanding of the degree to which the LCFS and RFS interact with one another and are complementary to one another.

The interactions of these two policies were tested using four pairwise scenarios. Each pairwise scenario consisted of an assumed RFS level (e.g., "No RFS," "Low RFS," and "High RFS") and LCFS level (e.g., "No LCFS" and "Base LCFS") with one pair incorporating a high crude oil price assumption (i.e., High RFS–High oil price). The "No Policy" scenarios show just what termination of the policy might cause—they are not showing the case of no past policies—and further adjustments could occur over time. Within each scenario we observe the complementary nature of the LCFS with respect to the RFS. Introducing the LCFS makes the RFS easier to meet, as evidenced by the lower RIN prices. At the same time, we observe a shift where biofuel use occurs. The LCFS incentivizes additional renewable fuel use in California at the expense of renewable fuel use in other states. Total renewable fuel use remains about the same. Additionally, we see a somewhat higher market share of mid- and high-blend fuels (e.g., E15 and E85) in California with the introduction of the LCFS. Broader agricultural commodity effects are muted within scenarios because there is little change in aggregate renewable fuel production. Thus the LCFS offers modest support to the RFS in these simulations.

Comparing across scenarios allows us to see the complementary nature of the RFS. As the RFS requirements rise we observe lower marginal LCFS credit prices. Although we do not see large changes in total motor fuel demand across scenarios, we do see increases in aggregate renewable fuel use. This also implies larger market shares for mid- and high-blend fuels regardless of region. In terms of commodity market effects, the increase in renewable fuel use across scenarios leads to additional demand for feedstocks such as corn for ethanol and soybean oil for biomass-based diesel. Higher prices for those commodities lead to acreage shifts that have positive cross-effects on the prices of other major commodities, such as wheat. These results suggest that the RFS can have important impacts on the LCFS.

It should be noted that we make several assumptions regarding the implementation of the renewable fuel policies in question. Most notable is the assumption requiring California to use renewable fuels in the order of their CI ranking, from the least carbon intense to the most carbon intense, and the assumption that obligated parties switch when 75% of that particular fuel has been consumed in California. While the first of these assumptions seems fairly logical, reflecting the likely choice of cost-minimizing firms, the second is somewhat arbitrary. If more low-CI fuels could be shipped to and used in California, or a smaller share, then the results could be affected.

Furthermore, this study does not account for other potentially related policy changes. At the time of writing, for example, there have been calls to extend the Reid Vapor Pressure waiver to include mid-level ethanol blends (specifically, E15), which would allow their sale year-round. This exemption might lead to a more rapid adoption of E15, making both the RFS and LCFS requirements somewhat easier to meet. While the specific results of such a change are beyond the scope of this study, it seems reasonable to think that the policy shift would not change the underlying interdependence of the RFS and LCFS although the magnitude of their interaction might be different.

References

California Air Resources Board, 2015. Final Regulation Order.
Christensen, A., Hobbs, B., 2016. A model of state and federal biofuel policy: feasibility assessment of the California low carbon fuel standard. Appl. Energy 169, 799–812.
Energy Information Administration, 2018. Annual Energy Outlook 2018: High Oil Price Case.
Environmental Protection Agency, 2017. Renewable fuel standard program: sandards for 2018 and biomass-based diesel volume for 2019. Fed. Regist. 82 (7), 58486–58527.

Huang, H., Khanna, M., Önal, H., Chen, X., 2013. Stacking low carbon policies on the renewable fuels standard: economic and greenhouse gas implications. Energy Policy 56, 5–15.

IHS Markit, 2016. Global Macroeconomic Outlook.

Rajagopal, D., 2015. On mitigating emissions leakage under biofuel policies. GCB Bioenergy 8, 471–480.

Rubin, J., Leiby, P., 2013. Tradable credits system design and cost savings for a national low carbon fuel standard for road transport. Energy Policy 56, 16–28. https://doi.org/10.1016/j.enpol.2012.05.031.

Rubin, J., Leiby, P., Brown, M., 2014. Regional credit trading: economic and greenhouse gas impacts of a national low carbon fuel standard. Trans. Res. Rec. J. Trans. Res. Board 2454, 28–35.

Thompson, W., Meyer, S., Westhoff, P., 2009. Renewable identification numbers (RINs) are the bell-wether of US biofuel mandates. EuroChoices.

Whistance, J., Thompson, W., 2014. Model documentation: US biofuels, corn processing, biomass-based diesel, and cellulosic biomass. FAPRI-MU Report #03-14.

Whistance, J., Thompson, W., Meyer, S., 2017. Interactions between California's low carbon fuel standard and the national renewable fuel standard. Energy Policy 101, 447–455.

Yang, Y., 2018. Improving estimates of subnational commodity flows in LCA for policy support: a US case study. Energy Policy 118, 312–316.

CHAPTER 11

Assessing the possibility of price-induced yield improvements to reduce land-use change emissions from ethanol

Jerome Dumortier*, Amani Elobeid[†], Miguel Carriquiry[‡]
[*]School of Public and Environmental Affairs, IUPUI, Indianapolis, IN, United States
[†]Department of Economics and Center for Agricultural and Rural Development, Iowa State University, Ames, IA, United States
[‡]Department of Economics, Institute of Economics, University of the Republic, Montevideo, Uruguay

Contents

1. Introduction

Over the last decade, there has been significant discussion about the economic and environmental consequences of corn ethanol production. Numerous studies have analyzed the effects on land-use change (Searchinger et al., 2008; Hertel et al., 2010; Dumortier et al., 2011; Carriquiry et al., 2019), commodity prices (Zilberman et al., 2013; Condon et al., 2015), and international trade (Elobeid and Tokgoz, 2008; Keeney and Hertel, 2009). The reason is the rapid increase in ethanol production and consumption prior to 2010 and the leveling off in the last few years (Fig. 11.1A). According to the US Department of Agriculture (USDA), 32.1% of the total corn supply in the 2016/17 marketing year was used for corn ethanol.[1] This increased use of corn as a feedstock for ethanol has implications on commodity prices and land-use allocation. The significant reallocation of land in 2007 between corn and soybeans is explained by expectations associated with the increase in corn prices in that time period

---

[1] USDA Economic Research Service Feed Grains: Yearbook Tables: https://www.ers.usda.gov/data-products/ feed-grains-database/feed-grains-yearbook-tables/.

Biofuels, Bioenergy and Food Security
https://doi.org/10.1016/B978-0-12-803954-0.00011-5

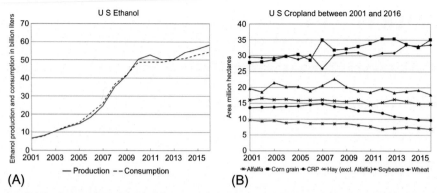

Fig. 11.1 (A) Evolution of US cropland for major crops between 2001 and 2016. (B) Ethanol production and consumption between 2001 and 2016. *(Data from (A) National Agricultural Statistics Service U.S. Department of Agriculture. (B) U.S. Energy Information Administration).*

(Fig. 11.1B). Since higher prices are not only observed by farmers in the United States but also globally, concerns about carbon release from land-use change are a significant part of the policy debate surrounding ethanol.

The effects on land-use change are of particular importance because to qualify as a renewable fuel, corn ethanol must achieve a reduction in lifecycle greenhouse gas (GHG) emissions by at least 20% compared to gasoline under the Energy Independence and Security Act of 2007. Initial lifecycle analysis (LCA) by Searchinger et al. (2008) and Fargione et al. (2008) has shown that corn ethanol could potentially increase GHG emissions compared to gasoline if indirect land-use change, that is, farmers in different parts of the world increasing crop acreage because of higher commodity prices, is taken into account. Subsequent research revises those initial estimates downward because market-mediating effects and price-induced yield increases can reduce those adverse land-use change effects (Hertel et al., 2010; Dumortier et al., 2011).

Wang et al. (2012) calculated the LCA emissions of five ethanol feedstocks, that is, corn, sugarcane, corn stover, switchgrass, and miscanthus, using the greenhouse gases, regulated emissions, and energy use in transportation model. Their results indicated that the lowest GHG savings (19%–48%) compared to gasoline are achieved with corn ethanol. A report by the USDA found that emissions from corn ethanol are about 43% lower compared to gasoline based on the recent performance of ethanol plants and the farm sector (USDA, 2017).

Dumortier et al. (2011) showed that land-use change emissions are very sensitive to assumptions with respect to crop yields. Small changes in yields can have large effects in terms of land-use change emissions because of the high carbon content in biomass and soil, especially in countries with tropical forests such as Brazil and Indonesia. Price-induced yield increases above the trend yield are possible due to higher commodity prices, that is, farmers find it profitable to change agricultural practices to increase yields on their cropland beyond what would have happened in the absence of higher prices. The argument of price-induced yield changes is difficult to measure and some research rejects this hypothesis (Roberts and Schlenker, 2009).

Related to the increase in yields is the question concerning the yield on newly converted cropland. The paper by Hertel et al. (2010) is based on the Global Trade Analysis Project and assumes that the yield on newly converted land is two-thirds compared to existing cropland. Taheripour et al. (2012) reassessed this assumption by using a global geographic information system dataset that specifies location-specific yields on new cropland. The authors showed that new cropland requirements decreased by 25% compared to the original two-thirds assumption. The analysis by Searchinger et al. (2008) assumed that any price-induced yield increase is offset by the lower yields on newly converted cropland.

Besides the questions whether price-induced yield increase is possible and how much yield is achievable on newly converted cropland, we are interested in how far farmers are currently from the yield frontier. Mueller et al. (2012) reported that significant increases (45%–70%) in production can be achieved for most crops by completely closing the yield gap. Producing on the yield frontier may be difficult to achieve and Foley et al. (2011) found that bringing yields within 95% of their potential yield for 16 crops could increase food production by 58%, and bringing them up to 75% would result in an increase in production by 28%.

In this chapter, we are interested in the yield growth that could potentially occur if countries move closer to the production frontier due to additional research or changing land management practices. Closing the yield gap for major crops globally would reduce the land requirements in case of an increase in biofuel production and thus would result in lower lifecycle GHG emissions. In general, LCA calculations are conducted by comparing a baseline to a scenario in which there is higher consumption/production of ethanol. Closing the yield gap would affect both the baseline and the scenario even in the absence of price-induced yield increases. The second relevant question is whether that yield increase occurs in regions that are carbon rich. Countries moving closer to their yield frontier in carbon-rich areas would be more advantageous in terms of avoided GHG emissions than for countries that are in a region with low soil and carbon biomass.

Many countries with large yield gaps have inadequate technological and economical resources (Grassini et al., 2013). To illustrate this point, we calculated the current and potential production for 12 major crops and for all countries based on yield gap data from Foley et al. (2011) and area data from Monfreda et al. (2008). We merge the results by crop and country with data on per capita gross domestic product (GDP) from the United Nations Statistics Division.[2] Fig. 11.2 shows this relationship between the yield gap and the per-capita GDP for the largest 50 producers per crop in terms of area. For commodities like barley, corn, sugar beet, rapeseed, and soybeans, a further increase in GDP does

---

[2] National Accounts Main Aggregates Database: https://unstats.un.org/unsd/snaama/selbasicFast.asp.

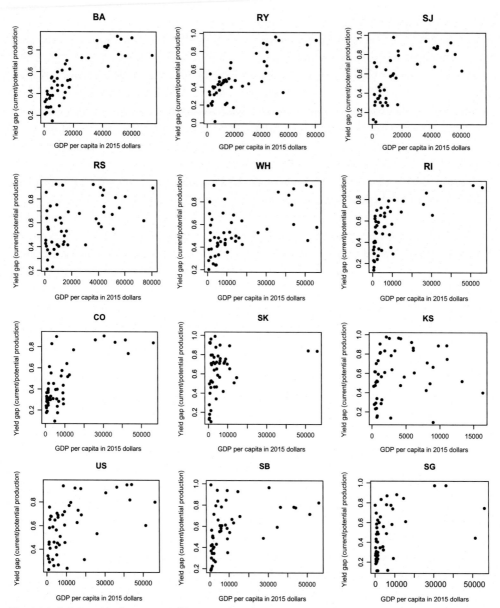

Fig. 11.2 Relationship between yield gap and per-capita gross domestic product for the 50 largest producers in terms of land area for 12 major crops.

not translate into a closing of the yield gap after a certain threshold is met in terms of per capita GDP.

In addition to taking into account the potential yield gap that exists for major crops, the modeling strategy for yield growth is also important because small changes can result in large differences in terms of area. Using a fixed growth rate of yields assumes that yields are increasing exponentially. Grassini et al. (2013) found that a linear trend (with declining growth rates over time) is more than adequate to describe the future yield for 36 countries and regions for corn, rice, and wheat production. In addition, yield growth will at some point plateau, that is, level off, due to the physical limitations imposed by soil and climate conditions. Grassini et al. (2013) concluded that "estimates of future crop production and land use must consider both historical yield trends and biophysical yield ceilings to improve forecasting capability." Grassini et al. (2013) were concerned about the possibility of yield plateaus for some major crops and countries such as rice in China or wheat in India and Northwest Europe.

Based on the fact that yield gaps exist and the importance of crop yields on evaluating the environmental performance of biofuels, this chapter addresses three issues. First, we identify the yield gaps for the crops and countries/regions covered in a global agricultural outlook model. The model has been used to assess LCA emissions from corn ethanol and includes price-induced yield increases. Second, we compare the yield gaps identified to the yield changes calculated by the model. We expect the largest percentage increases to occur in countries that are furthest away from the yield frontier. Third, we compare the yield gaps and the price-induced yield changes to the biomass content that would exist in the natural vegetation. If the yield gap is large in areas with a high biomass content in potential natural vegetation, then a policy focus may be warranted in those areas to reduce GHG emissions from an increase in biofuel production.

2. Modeling framework and data

We are combining two groups of datasets. First, we use data from the CARD/FAPRI Model to compare a baseline and a scenario. The baseline assumes status-quo policies and macroeconomic conditions. The scenario assumes that ethanol production in the United States is increasing by 15%. The model includes price-induced yield growth. These data are combined in a second step with information on yield gaps, land allocation, and carbon storage.

2.1 CARD/FAPRI model

The CARD/FAPRI Model[3] is a global partial-equilibrium model that forecasts agricultural production over the next 10–15 years. It covers 15 major crops and livestock categories for a total of 58 countries and regions. Smaller countries are grouped into regions, for example, other Asia, to achieve global coverage. The CARD/FAPRI Model was initially designed as a trade model able to predict the effects of policies on international

---

[3] Center for Agricultural and Rural Development (CARD) and Food and Agricultural Policy Research Institute (FAPRI).

agricultural trade. In recent years, the model has been used to predict land-use change to assess the effects of US biofuel policy. For our analysis, 2021/22 represents the final year of the predictions coinciding with the long-run equilibrium. That is, in 2021/22, all economic actors in the model make zero economic profit. The inputs to the model are policy parameters and macroeconomic projections such as economic growth and oil prices. The CARD/FAPRI Model then looks, based on input parameters such as yield, demand, and available area, for the commodity prices that clear the world market for all commodities. Note that the model is nonspatial in the sense that only world trade is calculated and not the trade flows between two countries. The output of the model is area, production, and consumption for the covered crop commodities as well as production and consumption for livestock.

Over the last decade, the model underwent significant updates to better incorporate land-use change triggered by biofuel policy. The land-use output of the model was first used by Searchinger et al. (2008) to assess GHG emissions from corn ethanol. Subsequently, the CARD/FAPRI Model was extended to include a GHG model (Dumortier et al., 2011, 2012), as well as a subnational Brazil model (Dumortier et al., 2012; Carriquiry et al., 2019). The GHG model quantifies the emissions from land-use change as well as from agricultural production (livestock and crop management). The subnational Brazil model was developed to better capture the dynamics of Brazilian agriculture at the regional level because expansion of cropland or pasture into the Amazon biome has important implications for global GHG emissions. The Brazil model also includes the pasture area based on cattle herd as output. A detailed description of the entire CARD/FAPRI Model and its components are available in Dumortier et al. (2011, 2012) and Elobeid et al. (2012).

2.2 Spatial information on agricultural production and carbon content

To assess our research questions, we complement the aforementioned CARD/FAPRI Model output from a 15% increase in ethanol production by three spatial datasets containing information on (1) carbon storage in natural vegetation, (2) yield gap by crop and country, and (3) spatial distribution of crops in terms of area. We have standardized all spatial datasets to a resolution of 0.5 degrees for each grid cell.

West et al. (2010) quantified the carbon trade-off between global crop production and natural vegetation. They calculated the ratio of change in carbon stock to crop yield. For example, in the subtropics, the average annual crop yield is 3.3 tons (t) ha^{-1} year^{-1} and the average change in carbon stock from land conversion is 68.3 t C year^{-1}. If compared to the tropics where the average annual crop yield is only 1.7 t ha^{-1} year^{-1} but the average change in carbon stock from land conversion is 120.3 t C year^{-1}, allocating cropland in the subtropics is more beneficial in terms of carbon emissions than in the tropics. Part of their assessment is a global map with the potential carbon content in natural vegetation.

The authors combine a map of natural vegetation with the Intergovernmental Panel on Climate Change Tier 1 method on carbon content. The resulting map contains the carbon content in potential natural vegetation at the global level. The dataset allows us to determine the cost of the yield gap for a particular crop and country with respect to carbon.

The key aspect of our analysis is the gap that exists between the observed yield and the potential yield. According to Foley et al. (2011), yield gaps result in different yields in areas that have the same growing conditions but different agricultural management practices. Foley et al. (2011) presented a global dataset that calculates the potential yield given current farming practices and technologies and compares those yields with the observed yield.[4] Their analysis is based on work by Licker et al. (2010) who determined the "climatic potential yield" for 18 major crops to the current yields. The analysis by Licker et al. (2010) was based on crops and yields in similar climatic regions. Foley et al. (2011) categorized the factors limiting yields in two categories, that is, nutrient limited and water limited. Both authors, Licker et al. (2010) and Foley et al. (2011), pointed out that closing the yield gap using conventional management practices such as irrigation and fertilizer may result in different environmental problems than the carbon release from requiring more land for the same amount of production. Their dataset is for the year 2000 and we calculate the yield gap and the ratio of current yield to potential yield. This implicitly assumes that the potential yield is increasing over time as well and that the gap remains constant over time.

The final spatial dataset contains information on the spatial distribution of crops at the global level (Monfreda et al., 2008). The dataset is compiled using national and subnational data sources and remote sensing information of cropland. The data reports the area harvested, yield, and production for the year 2000. The same dataset has also been used by Dumortier et al. (2012) to determine the effects of a potential expansion of global beef production. For our analysis, we assume that the fraction of land allocated to a particular crop in a grid cell is constant (compared to the overall crop production in a country).

3. Results

As aforementioned, the CARD/FAPRI Model was used to generate a baseline until 2021/22 to incorporate status-quo agricultural policies and macroeconomic forecasts. The model was then used to simulate a scenario that resulted in a 15% increase in ethanol production in the United States. In lifecycle emission analysis, the difference in land use and GHG emissions between the baseline and the scenario is attributed to the additional production of ethanol.

---

[4] The crops covered in the dataset by Foley et al. (2011) are barley, maize, palm oil, rapeseed, rice, rye, sorghum, soybeans, sugar beet, sugarcane, sunflower, and wheat. Their dataset does not include oats, cotton, and groundnuts/peanuts, which are covered in the CARD/FAPRI Model.

The 15% increase in ethanol production in the United States results in a total expansion of global crop area by 1.2 million ha representing an increase of 0.25% compared to the baseline. The largest relative increases can be observed in Mexico (0.76%), Brazil (0.52%), and the United States (0.44%). The increase in area in all other countries is 0.07%. The largest absolute increase occurs in the United States with an additional 399,176 ha followed by Russia (157,921 ha), Mexico (80,436 ha), Brazil (316,026 ha without the subnational model and 41,209 ha with the subnational model), and Indonesia (38,456 ha). All other countries combined increase their crop area by 208,536 ha. In addition, the scenario results in an increase in US prices for barley, corn, wheat, and soybeans by 2.4%. 3.4%, 1.5%, and 5.3%, respectively. Global prices for corn, soybeans, and wheat increase by 3.6%, 1.1%, and 1.3%, respectively. The total increase in ethanol production from the baseline to the scenario is 10.82 billion L in the United States.

Fig. 11.3 shows the yield increases or decreases that triggered a change in commodity prices between the baseline and the scenario. The results presented for the year 2021/22 correspond to the long-run economic equilibrium. For sugar, that is, sugar beets and

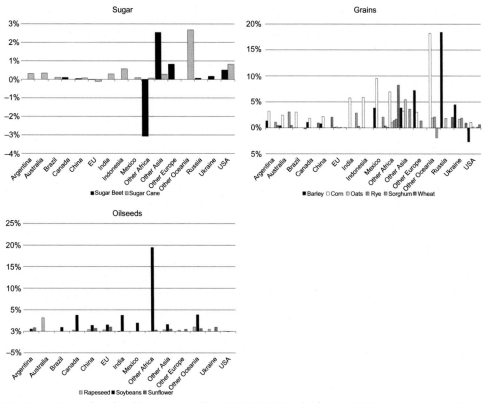

Fig. 11.3 Price-induced yield increase in the CARD/FAPRI Model from a 15% increase in US ethanol production.

sugarcane, the largest increases are observed in the United States (sugar beet and sugarcane), "Other Asia" (sugar beet and sugarcane), and "Other Oceania" (sugarcane only). Brazil as the largest producer of sugarcane increased yields by only 0.1%. In the category of grains, the most pronounced increases in yields occur for corn and soybeans because global prices increase the most for those commodities in the scenario. Note that significant gains are made in "Other Africa" and "India" for both corn and wheat as well as for oilseeds. In some cases, we observe a yield decrease compared to the baseline because we have a decrease in price. For grains, we see a significant increase in the regions "Other Africa" (especially corn and wheat) and "Other Asia" as well as in Russia (barley) and Mexico (corn). For oilseeds, the CARD/FAPRI Model predicts a significant increase in the yield of soybeans in "Other Africa."

As aforementioned, Grassini et al. (2013) suggested that the best fit to predict future yield was a linear trend that assumed a fixed gain in kilogram year^{-1}. Fig. 11.4 compares the yield growth assumed in the CARD/FAPRI Model over the projection period to the

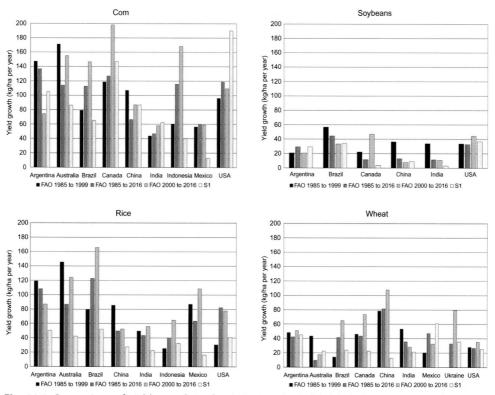

Fig. 11.4 Comparison of yield growth in the CARD/FAPRI Model (labeled "S1") over the projection period (until 2021/22) to the yield growth trends obtained from the Food and Agriculture Organization (FAO). The FAO yields are subdivided into three time periods indicated in the legend.

observed FAO yield growth for three time periods: (1) 1985–99, (2) 2000–16, and (3) 1985–2016. The yields presented assume a linear growth, that is, a declining growth rate over time. The first observation is the presence of significant variations in yield growth over time, crop, and country. For example, Argentina, Australia, and China experienced higher growth in the first time period (1985–99) than in the second time period (2000–16). For all four crops presented in Fig. 11.4, growth in the United States was smaller in the first time period than in the second time period. Except for wheat, China's yield growth in the second time period was smaller than in the first time period. Brazil as an important producer of agricultural commodities made advances in the production of corn, rice, and wheat but not soybeans for which it is a major producer. For large agricultural producers such as Argentina, China, Brazil, and the United States, the yield growth projections match historical observations fairly well except for wheat in China. Yield growth for corn in the United States is also above the long-term average.

The two main aspects of this chapter are the existing yield gaps and the carbon content of natural vegetation in countries that exhibit a yield gap. Fig. 11.5 illustrates the importance of a yield gap relative to the area of the largest agricultural producers in terms of area (15 largest producers in terms of area allocated to a particular crop are displayed). From an

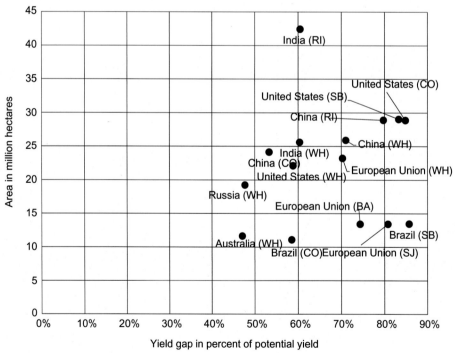

Fig. 11.5 Yield gap and area allocated for the 15 largest producers in terms of area.

environmental perspective, it would be desirable to have all countries situated on the right side, that is, closer to their potential yield, of the graph independent of the area devoted to a particular crop. The most "undesirable" location in the graph is the upper left corner since this would represent countries that are large in terms of crop area but are far from the potential yield. We can see that corn and soybeans in the United States as well as sugar beet in the European Union and soybeans in Brazil are all above 80% of their potential yield. Rice in India occupies the largest area for any crop/country combination and it only achieves 60% of its potential yield. Note that China is close to 80% for rice.

Indirect land-use change is a large component of lifecycle emissions from biofuels (USDA, 2017). Increasing yields in regions that have a high carbon content would reduce the land requirements in case of an increase in crop area due to biofuel policy. This is true even in the absence of price-induced yield improvements. To identify the areas that have large agricultural areas, yield gaps, and high carbon content, we constructed Figs. 11.6 and 11.7. Countries that have a high carbon content, a large yield gap, and are major producers would be of concern in terms of GHG emissions. For barley, Africa, Asia, Russia, and Ukraine are below 50% of their attainable yield but the carbon content in areas where barley is cultivated is generally small. A similar conclusion can be drawn for sorghum. Mexico, China, and countries in Africa plant large areas of corn in regions with a high biomass carbon content and a large yield gap. Policy interventions or technical assistance in those areas to close the yield gap would be beneficial in reducing GHG emissions from agriculture in general.

4. Conclusion

Over the last decade, there has been a large amount of research regarding the lifecycle emissions of corn ethanol and how those emissions are mitigated by a price-induced increase in yield. The price-induced increase in yield assumes that due to higher commodity prices, farmers are able to change management practices to achieve an above-trend increase in yields. Previous research assumes a constant elasticity of yield with respect to price, that is, a percentage increase in price results in an increase in yields. Environmental science and agronomy research show that there is a global variation in terms of countries being close to their yield potential. This yield potential is determined by the land and soil type as well as by the technological possibilities. Not all countries are close to their yield potential—especially low–income countries—and there is the potential to increase the yield in those countries given the price-induced yield increase. Other countries such as the United States, Brazil, and the European Union are already close to their yield potential for certain crops and thus it may be more difficult for those countries to increase the yield above the trend yield.

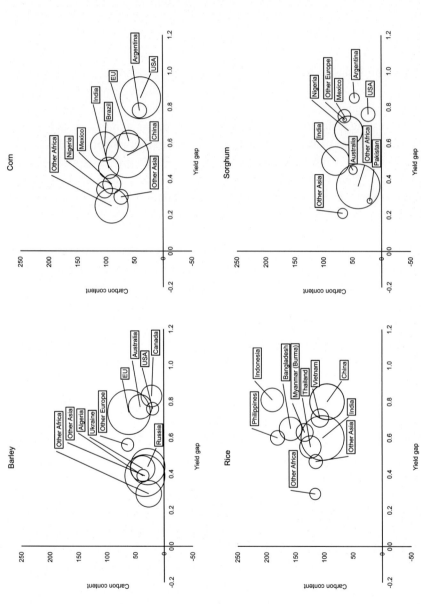

Fig. 11.6 Yield gap and carbon content per hectare in native vegetation for barley, corn, rice, and sorghum. The size of the circle represents the area in each country. Only the 10 largest countries in terms of area are selected for each graph.

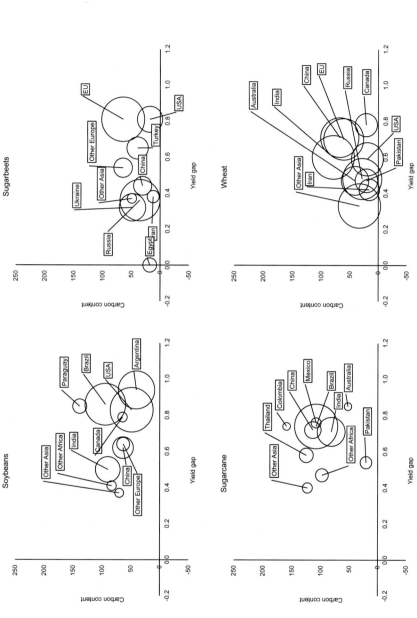

Fig. 11.7 Yield gap and carbon content per hectare in native vegetation for soybeans, sugar beet, sugarcane, and wheat. The size of the circle represents the area in each country. Only the 10 largest countries in terms of area are selected for each graph.

Note that the current analysis does not attempt to make the argument that the only possibility to increase yields above the trend is the production of ethanol since it could potentially lead to an increase in yields due to prices.

References

Carriquiry, M., Elobeid, A., Dumortier, J., Goodrich, R., 2019. Incorporating sub-national Brazilian agricultural production and land-use into U.S. biofuel policy evaluation. Appl. Econo. Perspect. Policy. https://doi.org/10.1093/aepp/ppy033.

Condon, N., Klemick, H., Wolverton, A., 2015. Impacts of ethanol policy on corn prices: a review and meta-analysis of recent evidence. Food Policy 51, 63–73.

Dumortier, J., Hayes, D.J., Carriquiry, M., Dong, F., Du, X., Elobeid, A., Fabiosa, J.F., Martin, P.A., Mulik, K., 2012. The effects of potential changes in United States beef production on global grazing systems and greenhouse gas emissions. Environ. Res. Lett. 7(2).

Dumortier, J., Hayes, D.J., Carriquiry, M., Dong, F., Du, X., Elobeid, A., Fabiosa, J.F., Tokgoz, S., 2011. Sensitivity of carbon emission estimates from indirect land-use change. Appl. Econo. Perspect. Policy 33 (3), 428–448.

Elobeid, A., Carriquiry, M., Fabiosa, J.F., 2012. Implications of global ethanol expansion on Brazilian regional land use. In: Socioeconomic and Environmental Impacts of Biofuels: Evidence From Developing Nations. Cambridge University Press, pp. 171–190.

Elobeid, A., Tokgoz, S., 2008. Removing distortions in the U.S. ethanol market: what does it imply for the United States and Brazil? Am. J. Agric. Econ. 90 (4), 918–932.

Fargione, J., Hill, J., Tilman, D., Polasky, S., Hawthorne, P., 2008. Land clearing and the biofuel carbon debt. Science 319 (5867), 1235–1238.

Foley, J.A., Ramankutty, N., Brauman, K.A., Cassidy, E.S., Gerber, J.S., Johnston, M., Mueller, N.D., O'Connell, C., Ray, D.K., West, P.C., Balzer, C., Bennett, E.M., Carpenter, S.R., Hill, J., Monfreda, C., Polasky, S., Rockstrom, J., Sheehan, J., Siebert, S., Tilman, D., Zaks, D.P.M., 2011. Solutions for a cultivated planet. Nature 478, 337–342.

Grassini, P., Eskridge, K.M., Cassman, K.G., 2013. Distinguishing between yield advances and yield plateaus in historical crop production trends. Nat. Commun. 4(2918).

Hertel, T.W., Golub, A.A., Jones, A.D., O'Hare, M., Plevin, R.J., Kammen, D.M., 2010. Effects of us maize ethanol on global land use and greenhouse gas emissions: estimating market-mediated responses. Bioscience 60 (3), 223–231.

Keeney, R., Hertel, T.W., 2009. The indirect land use impacts of United States biofuel policies: the importance of acreage, yield, and bilateral trade responses. Am. J. Agric. Econ. 91 (4), 895–909.

Licker, R., Johnston, M., Foley, J.A., Barford, C., Kucharik, C.J., Monfreda, C., Ramankutty, N., 2010. Mind the gap: how do climate and agricultural management explain the 'yield gap' of croplands around the world? Glob. Ecol. Biogeogr. 19 (6), 769–782.

Monfreda, C., Ramankutty, N., Foley, J.A., 2008. Farming the planet: 2. Geographic distribution of crop areas, yields, physiological types, and net primary production in they year 2000. Glob. Biogeochem. Cycles 22, 1–19.

Mueller, N.D., Gerber, J.S., Johnston, M., Ray, D.K., Ramankutty, N., Foley, J.A., 2012. Closing yield gaps through nutrient and water management. Nature 490, 254–257.

Roberts, M.J., Schlenker, W., 2009. World supply and demand of food commodity calories. Am. J. Agric. Econ. 91 (5), 1235–1242.

Searchinger, T., Heimlich, R., Houghton, R.A., Dong, F., Elobeid, A., Fabiosa, J., Tokgoz, S., Hayes, D., Yu, T.-H., 2008. Use of U.S. croplands for biofuels increases greenhouse gases through emissions from land-use change. Science 319 (5867), 1238–1240.

Taheripour, F., Zhuang, Q., Tyner, W.E., Lu, X., 2012. Biofuels, cropland expansion, and the extensive margin. Energy Sustain. Soc. 2 (25), 1–11.

USDA, 2017. A life-cycle analysis of the greenhouse gas emissions of corn-based ethanol. Tech. rep U.S. Department of Agriculture.

Wang, M., Han, J., Dunn, J.B., Cai, H., Elgowainy, A., 2012. Well-to-wheels energy use and greenhouse gas emissions of ethanol from corn, sugarcane and cellulosic biomass for US use. Environ. Res. Lett. 7, 045905.

West, P.C., Gibbs, H.K., Monfreda, C., Wagner, J., Barford, C.C., Carpenter, S.R., Foley, J.A., 2010. Trading carbon for food: global comparison of carbon stocks vs. crop yields on agricultural land. Proc. Natl. Acad. Sci. 107 (46), 19645–19648.

Zilberman, D., Hochman, G., Rajagopal, D., Sexton, S., Timilsina, G., 2013. The impact of biofuels on commodity food prices: assessment of findings. Am. J. Agric. Econ. 95 (2), 275–281.

SECTION 5

Institutional challenges and option

CHAPTER 12

Biofuels, food security, and sustainability

Parijat Ghosh, Patrick Westhoff, Deepayan Debnath
Food and Agricultural Policy Research Institute, University of Missouri, Columbia, MO, United States

Contents

1. Introduction

The initial progress of biofuels in industries was mainly driven by the inflation of fossil fuel prices. However, price and market forces cannot alone drive the whole process; intervention was needed in the form of policy and governance such as mandates and tariffs for import and keeping a focus on targeted domestic-based feedstocks like soybeans and maize. To make the process last, further research and studies on the development of feedstocks were needed. More versatile and higher yielding energy crops like sweet sorghum and jatropha were produced for next-generation biofuels. Although energy security pushed industries toward biofuels, a general concern could not be left unaddressed: climate change, which could turn the whole biofuel industry upside down. Increasing awareness of the relationship between the environment (in this case climate change)

Biofuels, Bioenergy and Food Security
https://doi.org/10.1016/B978-0-12-803954-0.00012-7

and sustainability has resulted in different views on biofuels, and how efficient biofuels are when using feedstocks to mitigate environmental impact. Moreover, are we willing to trade off between food and fuel? The food crisis of 2007–08 ushered a new debate and doubts about the sustainability of the biofuel system.

Linking the two aspects, food security, and biofuels, would require consideration of a few important conditions in the context of developing countries, for example, the industrial development of different countries, their different developmental model pathways, the capability of producers, and the agri-business conditions of these countries. However, the most important item that should be considered is the general state of food security of a particular country.

In addition to growing concerns of biofuels clashing with food security, it can be said that if sustainability is measured, then the feedstocks that are needed and grown for biofuels are not the same. Subsidies and mandates play a huge role in maintaining the biofuel market. The basic and established feedstocks for biodiesel (rapeseed, soybeans) and ethanol (maize) are thriving, but not much can be said about their future with respect to sustainability and economic growth. From the sustainability perspective, biofuels have both advantages and risks associated with them. The advantages are minimizing air pollution, greenhouse gas (GHG) emission reductions, and energy security, but the risks involve monocultures, the intensive use of resources, higher GHGs through land use change, and reduced biodiversity. To maximize benefits and to reduce risks, three pillars of sustainability—economy, ecology, and equity—have to be addressed. The future prospects of conventional biofuels need continuous innovation, research, strong support of the policy, and employment of efficient feedstocks. However, it should be noted that continued policy support can raise the question of competitiveness of the growing market of biofuels for the near future.

Keeping the surge of the biofuel market in mind and the growing concern of climate change, it becomes crucial to assess whether we want to trade, produce, or use biofuels sustainably. These concerns have arisen due to the perceived threat of impacting the environment negatively due to monoculture and GHG emissions through land-use change and deforestation. Therefore a shift is needed that could lead to a more sustainable energy system.

The following section briefly discusses the background of the theory of sustainability and the following section links this to biofuels. The last section discusses the implications of biofuels for food security.

2. Sustainability

Sustainability has been promoted as an essential condition for biofuels' long-term viability and for continued public support for renewable energy and climate change mitigation.

2.1 Emergence of the concept of sustainability

While the definition of sustainability is contested, there is a basic understanding of sustainability as a symbiotic relationship between human beings and nature that meets the needs of the present generation, while securing the needs of future generations without compromise and keeping a harmonious relationship with the environment and human beings. "Sustainable development is a development that meets the needs of the present without compromising the ability of future generations to meet their own needs" (Brundtland Report; WCED, 1987:43). Sustainability has been acknowledged for three centuries but with different approaches and meanings in German, French, and Dutch. It then found its place in the 20th century as a new evolving concern in English (Pisani, 2006). The concept of sustainable development is multidisciplinary. It emerged with the growing awareness of an ecological crisis around the end of the 20th century. In the world of environmental discourse, the words "sustainable development" have become a central theme. However, they have also become buzzwords and are overworked. As a result, "sustainable development" challenges the pedagogy in various bodies of science. The words are often used without giving much thought to their real meaning and implications in the correct context. Therefore it is important to know what caused the concept to emerge and the importance of its implication for development.

The concept of sustainability is multidimensional and includes nature, society, and economy, that is, it does not limit itself just to the environment (Walsh, 2011). The idea of sustainability to meet the demands of the present and future could only be possible if there was a constant flow of capital into society and communities (Alberti, 1996; Callaghan and Colton, 2008). Sustainable development should focus on the quality of life with lasting satisfaction. It is also linked to equity, which allows humans to achieve their needs while keeping the triple bottom line: society, environment, and economy (Brundtland Report, 1987).

The last decade has seen a constant disagreement with the definition and the significance of the term among individuals in various professions (Morelli, 2011). Two schools of thought have emerged from this debate: the first focuses on the three pillars of sustainability—economy, ecology, and society—simultaneously benefiting from each other; the second views sustainable development as a synergetic relationship between nature and human society (Robinson, 2004). Although no consensus has emerged, the debate has resulted in a concept that is more open to be interpreted by individuals professionally and politically rather than logically or scientifically (Robinson, 2004). In the business world, progress is even more limited (Morelli, 2011). Some industry sectors have drilled down into sustainability, however. For instance, sustainable agriculture focuses on site-specific farming practices that aim at meeting the demand for present and future food, fiber, and environmental and energy services; goals include soil conservation, enhancing biodiversity and clean water, and improving the quality of life of farmers and consumers. In sustainable agriculture, it is important to maintain a balance between ecology, profitability, and energy efficiency (Menalled et al., 2008). Morelli (2011) in his work stated that the concept of sustainability if treated as a concept by itself is much more discredited than when it is being used as a delineating modifier such as the economy, agriculture, or ecology. A constant effort has been made by members of various professions to define the three pillars of sustainability and the concept of sustainability within the context of their own respective professions.

Although the terms "sustainability" and "sustainable" appeared in the *Oxford English Dictionary* for the first time during the second half of the 20th century, the equivalent term has existed for centuries in French (durable), German (lastingness), and Dutch (duurzaam). Ecological problems like salinization, deforestation, and infertility of soil or soil erosion occurred as far back as ancient Egyptian, Greek, Mesopotamian, and Roman civilizations, which today we term as sustainability problems. Environmental degradation was caused by early civilizations due to human activities, such as farming, mining, and logging, and there became a need to implement sustainable practices to maintain the "everlasting youth" of the earth (Columella, 1948; Pisani, 2006).

Up to at least the 18th century, wood was used both as fuel and material for construction, and soon it became overutilized. Georg Agricola (1950) mentioned the negative impacts of woodcutting in an extensive manner and that it could threaten human existence. The shortage of wood stimulated human beings to think in a different way in favor of using natural resources in a responsible manner for present and future generations by promoting "sustainable development" (Pisani, 2006). To add to this concern, there was another growing issue—the ever-increasing population. Thomas Robert Malthus, in his work *Essay on the Principle of Population as it Affects the Future Improvement of Society*, published in 1798, stated that the increase in population needs to be controlled because it will threaten to outstrip food production.

In the 19th century, coal became the source of energy and concerns were raised that overmining of coal may exhaust the resource. Should the wasteful use of coal not be stopped then England may lose her position in the industry, as was written by Jevons (1866) in his work *The Coal Question*. For more than a century there were different remarkable works that dealt with the same concept, which we call today "sustainable development." In *Principles of Political Economy*, published in 1848, John Stuart Mill in his short chapter "Stationary state" stated that capital and population can be static but not human development. "I sincerely hope, for the sake of posterity" that the world's population "will be content to be stationary, long before necessity compels them to it" (Pisani, 2006). Similarly, other scholars like Marsh 1965, writing along the same line, talked about environmental degradation, disturbed by human actions and intervention, which may lead the earth to be a place unfit for human habitation, and the possible remedies that need to be undertaken for protecting the environment.

After wood and coal in the 20th century, oil became the primary source of energy, which again with its overuse raised concerns that it may be exhausted. To ensure that human beings can live in a civilized world, it is important that they use natural resources in a responsible manner. From the foregoing literature, it is clear that the concept of sustainability started in ancient times with the overconsumption of wood, coal, and oil together with the increase in the human population. This boosted the need for people to be responsible for their actions to help maintain the living standards of future generations.

2.2 20th century sustainability

The 20th century saw a drastic change in terms of production, consumption, and accumulation of wealth, mainly after the industrial revolution. With the increase in population, there was a need for unlimited economic growth with regard to human development. Scientific and technological advances gave hope to look for unlimited possibilities in the search for alternative forms of energy. The expectation for ever-increasing affluence created by industrial and commercial expansion gave a new turn to the human problem, that is, environmental crisis. Economists from the neoclassical orthodox school were much more aware of related sustainability problems caused by overconsumption and exploitation of natural resources. However, they were assuming that with scientific and technological advances they can overcome such crisis by economizing the scarce input (Pisani, 2006). Nevertheless, scientific and technological advancement was also the cause of untold damage to the environment and natural resources. This ecological crisis became more acute and radical because the future of the human race was also called into question.

Some of the prominent theories that shaped the concept of "development" with respect to developed and developing countries were initiated in the colonial period. Modernization theory and dependency theory are the dominant concepts that offered different solutions to promote growth rates and the expansion of world markets. These basic theories played a big role in the changing situation, especially development discourses and sustainable development.

Expectations of creating more wealth and unlimited economic growth came to a halt because of the worldwide recession between 1974 and 1976, which led to an awareness of the limits to economic growth. People became skeptical about the blissful effects of scientific and technological promises.

According to the Brundtland Report, and in a broad sense the definition of sustainable development, economic, ecological, and social aspects are supposed to have equal weight as shown in Fig. 12.1. Considering their applications, economic development was different in different centuries. Research conducted over the years attempted to bring the other two aspects back into balance, but these efforts are still awaiting positive outcomes (Fig. 12.2).

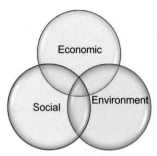

Fig. 12.1 Sustainable development: the three pillars of sustainability.

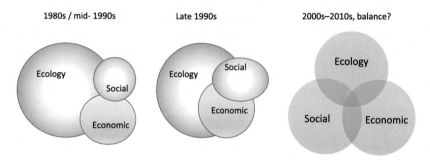

Adopted from: Dr. Andrea Colantonio, Oxford Institute for Sustainable Development (OISD), 2009.

Fig. 12.2 Evaluation in the weight of the sustainability matrix over time. *(Adapted from Colantonio, A. 2009. "Measuring Social Sustainability: Best Practice from Urban Renewal in the EU." Oxford Institute for Sustainable Development (OISD) - International Land Markets Group Originally from Marghescu, T. (2005), 'Greening the Lisbon Agenda? = Greenwashing?' Presentation at the Greening of The Lisbon Agenda Conference, EPSD, 23 February 2005 European Parliament, Strasbourg.).*

2.3 Bioenergy and sustainable development

This roadmap concentrates on identifying opportunities to produce and use bioenergy sustainably, that is, in ways that avoid negative impacts on the environment, that foster both food and energy security, and contribute to sustainable development goals for agriculture, rural development, and climate. Like other renewable energy technologies, bioenergy can provide a number of environmental and social benefits. It can:

- Reduce GHG emissions (especially in sectors such as long–haul transport where other opportunities are limited).
- Improve energy security by enhancing the diversity of energy supply and reducing exposure to fluctuating global energy markets and import dependency.
- Provide economic opportunities, including jobs and income for rural economies.
- Complement efforts to improve waste management and air and water quality.
- Contribute to the improvement of modern energy access for heating, cooking, and electricity for the 2.7 billion people who lack it.

- Support investments in rural infrastructure and development that are essential for improving food security.
- Provide additional market incentives and opportunities for afforestation and reclamation of degraded lands.

Bioenergy interacts extensively with the agricultural, forestry, and waste management sectors. The related environmental, economic, and social implications associated with the production and use of bioenergy have many implications that reach beyond the energy sector. These create both benefits and potential risks for the environmental, social, and economic pillars of sustainability.

These benefits and potential risks relate particularly to the United Nations Sustainable Development Goals (SDGs). An ongoing analysis being conducted as part of the Global Bioenergy Partnership activities notes that while biomass, bioenergy, and biofuels are not explicitly mentioned in the SDGs, bioenergy has the potential to contribute to or have positive impacts on nearly all the SDGs (FAO, 2011).

The SDGs can drive the expanded use of bioenergy as part of a growing bioeconomy, while also providing safeguards against unsustainable bioenergy practices. This goes beyond SDG 7, which is primarily concerned with energy. For example, bioenergy can contribute to combatting climate change (SDG 13). SDG 3 (Health) can be a driver for avoiding the health implications of air pollution due to the inefficient traditional use of biomass, while encouraging the efficient use of biomass to replace polluting fossil fuels.

3. Biofuels and the sustainability challenge

The concept of sustainability is complex and multidimensional. To implement the goals of sustainability in a pragmatic way a greater understanding is required in a specific local context. To meet the sustainable development goal, sustainable biofuel production needs to address the three pillars of sustainability—one that is economically viable, conserves the natural resources addressing environmental protection, and ensures the well-being of individuals and society. Moreover, the three pillars of sustainability—economic, ecological, and social—could be best addressed if they are taken together in a holistic manner rather than treating them individually.

To measure biofuels in the framework of sustainability offers both positive and negative impacts. The positive side of biofuels increases energy security, improves the quality of air, spurs the quality of city life, contributes to the growth of rural life, and reduces GHG emissions. On the downside, biofuels under intensive production systems can impact negatively on biodiversity such as monoculture and deforestation, and conflict with issues such as the production of food and food security, the availability of water where there is water scarcity, water quality, soil degradation, negative carbon release, as well as increasing GHG emissions due to indirect land-use change.

It is always challenging to balance the three dimensions of sustainability: economic benefits with environmental and social impacts. In this case also, even if biofuels partially meet the ecological and social aspects, they need to pass the economic aspect too. This means that economic viability is a must and it should include efficiency of production with high yields, ensuring long-term benefits and profits, intensive management, access to resources (land and labor) and advanced technology, and keeping the market competitive and growing. To address all these simultaneously is a delicate act and needs much understanding and knowledge.

3.1 Economic sustainability

The economics of biofuels is driven by long-term profitability and market competitiveness with fossil fuels, which is possible because of active policy support like subsidies and mandates. In general, economic sustainability requires profit-driven strategies and stability. The support of policy measures makes it difficult to assess the future of biofuel systems in terms of economic viability. However, because of technological improvements (market opportunities and use of by-products and diversification, efficient consumption of internal energy, etc.) along with the support and protection of the biofuel industry domestically (Brazil's sugarcane ethanol in the 1970s, corn–ethanol from the United States, and EU rapeseed-based biodiesel) long-run costs have been managed and economies of scales have been developed.

The food-versus-fuel debate was triggered by the food crisis in 2007–08, which raised a concern: should we compromise between food production over the expansion of biofuels? Expansion of biofuels is needed to meet the ever-rising demand for energy. This expansion if not controlled may push food production into the margins, which could result in lower yields. Not only that, other resources that are needed to produce both food crops and bioenergy crops like water and fertilizers may constrain food availability. The yield of food production could also see growth if there is a hike in the market price of land and adaptation of improved technologies, which can boost productivity through intermixed cropping and rotation. First-generation biofuels that are slow in adopting advanced and progressive technology could improve energy input/output ratios and increase market value by advancing the use of by-products. However, these effects depend much on location and local market conditions along with the support of policies.

Agriculture commodity prices tend to rise due to an increase in demand for feedstocks from the biofuels sector. Also, trade plays a big role in globalizing the biofuels market. On the other hand, if biofuels are more openly traded by allowing the trade barriers to be lowered, which includes tariffs, then there could be considerable moderation of overall agricultural commodity prices. The competition between biofuels and food production over shared resources depends on the hike in commodity prices making food more expensive, which demands resources to be allocated to food production from biofuels. This competition will make it difficult to access the financial viability and sustainability of biofuels in the near future and in the long run, because linkages between energy and food would grow stronger with time. It is still an uncertainty if biofuels in their commercial form would be available to the second generation; if yes, then the competition would become more intense with respect to shared productive resources. At this stage, there would be a need for policies and regulations to step in and balance the two extremes: food and energy security.

3.2 Environmental sustainability

Environmental sustainability covers a wide range of issues starting from a specific location to global. Global issues comprise concerns about GHG mitigation, climate change, and renewable energy, while the location-specific issues are soil erosion, water management, soil quality, and air and water pollution. The role of biofuel in the dimension of environmental sustainability is largely to reduce GHG emissions (e.g., CO_2, methane, N_2O), though there are controversies regarding its effectiveness. The leading sources of GHG emissions for non-CO_2 GHGs are agricultural practices like the use of fertilizer, soil tillage, pesticides, irrigation practices, and harvesting. In evaluating the environmental factor, the use of land prior to the production of biofuel plays a significant role. If forest or grassland are used for the conversion of biofuel, then the reduction of GHG emissions is markedly affected. Sustainability of biomass-based biofuel is increasingly measured via lifecycle analyses (LCAs). All the phases—input and output data (biomass production, feedstock storage and transportation, biofuel production, and transportation and final use)—of the product's lifecycle are required in an LCA. All output data are accounted for, including both leaked and captured gases and by-products. However, in an LCA the results may vary considerably regarding the environmental impact of biofuel production and consumption because it mostly depends on the assumption of the main parameters.

Biofuel can replace fossil fuel through energy substitution, which could be another important motivation for the use of biofuels. It is important to track the energy content of the biofuel as well as the amount of fossil fuel used during its production, which will decide biofuels' contribution to reducing GHG emissions. The energy requirement in the production of biofuel includes the energy needed for both cultivation and harvesting the feedstock. In the cultivation process, the energy required is in the form of pesticides, fertilizers, tillage, and irrigation technology. Also, it is required to process the feedstock into biofuel and transport it through both the production and distribution processes.

The ratio between energy output and the input of fossil energy needed for production of the resultant biofuel is defined as the fossil energy balance. This plays a significant role in judging the desired biofuel derived from biomass. It explains to what extent biomass could replace fossil fuel. Fig. 12.3 shows reported theoretical ranges of fossil energy balances of liquid biofuels according to fuel and feedstock.

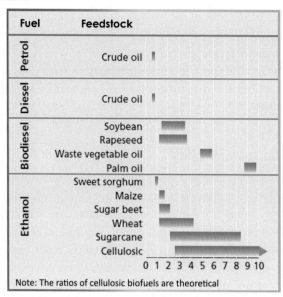

Fig. 12.3 Fossil energy balances of liquid biofuels according to fuel and feedstock. *(Courtesy: FAO (Food and Agricultural Organization of the United Nations) (2008). The State of Food and Agriculture 2008).*

The energy balance for conventional diesel and petrol ranges between 0.8 and 0.9. This is because some of the energy is consumed in the process of refining crude oil into petroleum and diesel; also, energy is consumed transporting it to the market. Now, to use biofuel as a replacement for conventional fuels, the energy balance of the biofuel should exceed 0.8–0.9. For sugarcane-based ethanol, the estimated balance ranges from 2 to 8, while for maize-based ethanol, it is 1.34. However, cellulosic ethanol has the highest fossil energy balance ranging between 3 and 10. Even though the ranges of those feedstock specific energy balance varies across studies the overall conclusion remain same.

The other consideration besides GHG and energy is water management. In the context of evaluating environmental sustainability, it becomes important to consider the quantity of water used and its impact on local water quality and availability for the future. Managing fertilizer runoff deriving from agricultural land to nearby streams and rivers is another challenge.

Avoiding biodiversity loss that takes place because biofuel is another criterion to maintain sustainability. Preserving biodiversity is very important, but the production of biomass under an intensive monoculture system may lead to loss of biodiversity. Not only that, it may have a negative impact on habitat loss, contamination from herbicides and fertilizers, and expansion of invasive species. However, on the positive side, biomass deployment from degraded land may have a beneficial effect.

3.3 Social sustainability

The social pillar of biofuel sustainability relates to the development of rural areas with inclusive growth and reduction in poverty. The ownership of land and labor rights supports the social aspects of sustainability.

The appropriate use of land is very much needed because it is a limited resource and much of it use depends on the owner. The value of land can be determined in multiple ways, such as conservation, generation of wealth, and ecosystem servicing. The role of biofuel here is that it enhances commercial opportunities and increases the contribution of land, benefiting all as a whole.

With respect to rural development, biofuel production has offered opportunities for employment. Especially in developing countries, biofuel production has acted as a big driver by stimulating employment. The use of new and advanced technology plays a big role in attracting capital to the agricultural sector and increasing productivity with access to better fertilizers, high yielding varieties, and infrastructure. Biofuel production also has a positive impact on human welfare by increasing access to energy services. In rural areas, women and children are mostly responsible for collecting firewood, and biofuel production make it possible for better access to pumped portable water and electricity, reducing their workload and supporting a better life. This all shows how increased opportunity for employment and higher wages in rural areas could have a positive impact on individuals' lives and a positive effect on the local economy.

4. Food security

4.1 Concept of food security

To evaluate the impact of biofuels on food security requires a working definition of food security. In its 2017 report on the state of food insecurity (http://www.fao.org/3/a-I7695e.pdf), the FAO stated that food security is "a situation that exists when all people, at all times, have physical, social and economic access to sufficient, safe and nutritious food that meets their dietary needs and food preferences for an active and healthy life."

The FAO definition suggests that food security means a lot more than simply supplying enough food in an average year to meet the population's caloric needs. Food supply is important, of course, but all people have to be able to obtain the food they need to live a healthy life. A country is not food secure if certain people lack access to nutritious food because it is unavailable or unaffordable.

Note that this definition of food security does not require reliance on local food production. A country that imports much of its food supply may be more food secure than a country that imports no food but experiences spikes in food prices when drought reduces production of a staple crop. Some countries impose tariffs in pursuit of food security, and it is true that tariffs can reduce food imports, thus making a country appear more self-sufficient. However, tariffs also can also increase food prices, making it harder for low-income consumers to obtain the food they need. Therefore imposing a tariff on food products can reduce food security, as defined by FAO, even though it reduces food imports.

FAO goes on to identify four dimensions of food security. *Food availability*, of course, is important, and one would expect that efforts to increase the supply of healthy food would improve food security. *Economic and physical access to food* is equally critical; a country with bountiful food supplies can still have many people who are food insecure because they cannot afford to buy healthy food. Poor health, unsafe water, and a lack of micronutrients can interfere with *food utilization*. Finally, *stability* is also required; a person is not food secure if they cannot get the food they need when they lose their job, become sick, or experience a crop failure.

4.2 Maize-based ethanol and the food-versus-fuel debate

Biofuels affect food security in many ways. The case of maize-based ethanol in the United States is an interesting case study that illustrates how complex the story can be. Between 2005 and 2010, the United States tripled its production and use of ethanol because of supportive policies and a sharp increase in petroleum prices (EIA, 2018). Over the same period, world grain prices also increased sharply, with US export prices for maize increasing from $106 per ton for the crops harvested in 2005, to $277 per ton for the crops harvested in 2010 (https://data.ers.usda.gov/FEED-GRAINS-custom-query.aspx).

The increase in the use of maize to make ethanol was one factor contributing to the increase in grain prices. When ethanol plants used more grain to make fuel, less was available to produce food for people and feed for livestock, and this pushed grain prices higher. The food-versus-fuel debate intensified as food prices rose. UN special rapporteur on the right to food Jean Ziegler said that it was a "crime against humanity" to divert arable land to the production of crops that would be used to produce fuel instead of food (http://news.bbc.co.uk/2/hi/7065061.stm).

A number of studies (Abbott et al., 2008; Taheripour and Tyner, 2013; Westhoff, 2010; Hochman et al., 2014; de Gorter et al., 2013) tried to identify the causes of the increase in grain prices during the biofuel boom. In addition to the increase in biofuel production, a number of other factors also contributed, including weather-induced crop shortfalls, dietary changes in middle-income countries, exchange rate movements, and changes in farm and food policies. While the various studies largely agreed on the list of factors, they differed in their assessment of the magnitude of the effects.

Between the 2005 and 2010 marketing years, the amount of US maize used in ethanol plants increased by 87 million metric tons (3.4 billion bushels). It is easy to portray this as a huge change and conclude that it must have had a large impact on grain prices (USDA, 2018). The tripling of maize used in US ethanol plants coincided with a significant reduction in US maize exports and reduced use of maize grain in US livestock feed rations. The increase in maize used in US ethanol plants accounted for more than half in the 149 million-ton increase in world maize use, even as the world population grew and China increased its demand for meat and dairy products produced from maize-consuming livestock. US and global stocks of maize declined relative to consumption, an indicator that normally corresponds with rising prices.

Biofuel supporters highlighted a different set of facts to suggest the effect of increased US ethanol production on world food prices was small. In 2010, the total US use of maize by ethanol plants was 127 million metric tons, less than 15% of global maize use. Furthermore, for every ton of maize that enters an ethanol plant, approximately 300 kg of dried distillers grains are generated as a coproduct when fuel ethanol is produced. Distillers grains can replace maize and other feeds in livestock rations, so the net effect of ethanol production on food and feed markets is less than would be suggested by the gross amount of maize used by ethanol plants. Finally, farmers around the world expanded production in response to higher prices. Global maize production increased by 19% between 2005 and 2010, with increases in both the area devoted to maize production and in yields per hectare (USDA, 2018).

Even if it is unclear what proportion of the increase in maize prices was due to the growth in US ethanol production, higher maize prices clearly have implications for food security. Millions of people in Latin America and Africa receive a large share of their daily caloric intake from tortillas, maize meal, and other foods made directly from maize. Higher maize prices increase the cost of those foods, making it more difficult for low-income families to meet their nutritional needs. Maize is also the most important livestock feed in the world. Higher maize prices increase feed costs, and these higher costs reduce animal protein production. Reduced supplies increase consumer-level prices for meat, milk, and eggs, with implications for dietary choices. Even in a high-income country like the United States, higher consumer meat prices contributed to a 9% reduction in total beef, pork, and poultry consumption between 2007 and 2012 (USDA, 2018).

Higher maize prices also have impacts on other crops. Part of the increase in the maize area came at the expense of other crops. Higher maize prices also encourage livestock feeders, food processors, and consumers to reduce their use of maize and increase their use of other grains. These supply and demand substitution effects mean that higher maize prices also correspond with higher prices for wheat, rice, soybeans, and other crops. Thus an increase in US ethanol production can have indirect effects on the cost of food for someone who consumes no maize or animal protein.

For a low-income urban family that finds it difficult to obtain the food it needs, higher maize prices are almost certain to have a negative effect on food security. However, the story can be more complicated for other groups. A poor farm family that sells more grain than it produces will experience an increase in income when grain prices rise. That increase in income can result in a more nutritious diet, suggesting that increased US ethanol production may have improved the food security of some rural households around the world. Higher grain prices and production could also increase employment and wages paid to farmworkers. Thus for some rural households, the net effect on food security of increased US ethanol production may depend on the balance of the impacts on food prices and family income.

While undernutrition continues to be a central food security concern, overnutrition is also a growing problem, as rates of obesity and associated health problems increase. If higher maize prices raise the price of foods high in calories or fat, it may encourage healthier diets. Higher grain prices may simultaneously increase problems with undernu-trition but reduce problems related to overnutrition.

Higher grain prices in international markets do not translate into higher prices in every village in every country. This can happen when government policies restrict trade or when high transportation costs effectively isolate local markets from developments in international trade. Some of the areas in the world with the highest levels of food inse-curity are also the areas most isolated from global markets. These areas will not be affected positively or negatively when US maize ethanol production increases the prices of grain traded in international markets.

Given all of these offsetting considerations, what was the net effect of increased US ethanol production on global food security? FAO (http://www.fao.org/3/a-I7695e.pdf, page 5) reports that the number of people undernourished in the world declined from 926 million (14.2% of the population) to 795 million (11.5%) between 2005 and 2010. That number leveled off in subsequent years, even as the growth in US ethanol production slowed and global grain prices retreated from their peak levels. This is not proof that US increased ethanol production improved global food security, but rather a reminder that many factors affect food availability and access.

4.3 A broader consideration of biofuels and food security

As the US maize ethanol story makes clear, the relationship between biofuel production and food security is complex. Other biofuels have characteristics that may result in qual-itatively different impacts. Like maize ethanol, some biofuels use food crops as feedstocks; others do not. Some biofuel feedstocks grow on vast tracks of land that compete directly with food production; other biofuels use waste materials and have little impact on food availability. Biofuels production costs vary greatly, with some biofuels requiring strict mandates or heavy subsidies, while others might continue to be produced even if sup-portive policies were removed.

Before the United States greatly expanded its production of maize ethanol, Brazil was the top producer of ethanol in the world, with sugarcane the dominant feedstock. Like maize-based ethanol, sugarcane-based ethanol, directly and indirectly, reduces food availability. Sugarcane that could have been used to produce sugar is instead converted into ethanol, and the land used to produce the feedstock is not available to grow other crops. The result is higher prices for sugar and for other crops.

Sugar's contribution to a healthy diet is dubious, but sugar provides a significant share of total caloric intake for much of the world's population, including people who struggle to consume enough calories during the course of a day. When sugar prices increase, the cost of feeding a poor family increases, which can result in increased undernourishment. Of course, higher sugar prices discourage sugar consumption, which may provide health benefits, especially to those who are overweight or obese.

Sugarcane production provides income to farmers and employment to farm workers. Ethanol production plants also create jobs and economic activity. As with maize ethanol, the increases in household income for some families must be balanced against the increase in food costs when evaluating impacts on food security. Historically, sugarcane production in Brazil was much more labor intensive than grain production in the United States, but mechanization and other efficiency gains have reduced the number of jobs created from each million liters of sugar-based ethanol production.

The relative economic costs of maize-based ethanol and sugar-based ethanol have changed over time. It was once conventional wisdom that it was more economically efficient to produce ethanol from sugarcane than from maize, but production technologies have changed and the market price of US maize ethanol has often dropped below the price of Brazilian sugarcane ethanol in recent years. The two types of ethanol have different environmental implications, and much LCA has concluded that sugar-based ethanol does more to reduce GHGs and has other environmental advantages relative to maize ethanol (Debnath et al., 2017; EPA, 2018; Broch et al., 2013; Searchinger et al., 2008; Overmars et al., 2015). If these assessments are correct, sugarcane-based ethanol may have an advantage over maize-based ethanol in terms of long-term sustainability, with implications for food security as well.

Besides ethanol made from maize or sugarcane, the other main biofuel produced in the world is biodiesel made from vegetable oils. Various feedstocks are used, including rapeseed oil, soybean oil, and palm oil, each with its own agronomic characteristics and implications for food security. When palm oil from Malaysia or Indonesia is used to produce biodiesel, it increases the price of palm oil. As with other food-based biofuels, this increases consumer food costs but also increases incomes for those associated with the industry. Higher palm oil prices result in an expansion of palm plantations, which displace other food crops and forest uses of land, with implications for both food security and the environment.

Biodiesel made from oilseeds like rapeseed and soybeans is distinct in an important respect from the other biofuels discussed so far. Oilseeds are crushed to produce both vegetable oil and protein meal, but only the vegetable oil is converted into biodiesel. Soybeans, in particular, produce a lot more protein meal than vegetable oil when crushed. When biodiesel production increases the demand for and prices of vegetable oils, it also raises prices of the oilseeds. Higher prices for soybeans and rapeseed encourage farmers to plant more of those crops, which in turn increases production not only of vegetable oil but also of protein meal. Depending on other market interactions, this can actually result in lower prices for oilseed meals.

Oilseed meals provide protein in livestock feed rations. If oilseed meal prices fall but other feed prices remain unchanged, the result of an increase in biodiesel production can be a reduction in livestock and poultry feed costs. This effect could be especially pronounced for poultry and pork production, where feed rations include a high proportion of oilseed meal. Lower feed costs, in turn, could result in increased production of poultry and pork, and lower meat and egg prices to consumers. Thus, unlike corn- or sugar-based ethanol, it is at least plausible that oilseed-based biodiesel could reduce consumer costs for some foods, offsetting some of the effects of higher consumer prices for vegetable oil. This result is far from certain, however; if an expansion of oilseed production comes at the expanse of grain production, the resulting increase in grain prices could offset part or all of the reduced feed ration costs resulting from increased oilseed meal production.

Of course, biofuels can also be made from feedstocks not commonly used in food production. For example, research has shown that it is technically possible to make ethanol from switchgrass, miscanthus, and other warm season grasses not regularly consumed by people or even livestock. Many people support these types of biofuels because they appear to avoid the food-versus-fuel debate. However, there are at least two issues to consider. First, biofuel production from these feedstocks has not been profitable, as per-liter costs have exceeded those for conventional biofuels given currently available technologies. Second, even though these feedstocks do not compete directly with human food, there may be important indirect effects. Land used to produce switchgrass cannot simultaneously be used to produce grain or even livestock forage. Some argue that such crops could be grown on marginal land not utilized to produce food, but there is little evidence that such production is likely to be economically feasible.

Making biofuels from waste products like vegetable oil previously used to prepare fried foods in restaurants may be an attractive alternative. To the extent such feedstocks are truly waste products that would otherwise be discarded, the result can be all of the benefits of biofuel production without any of the negative impacts on food security discussed here. The amount of biofuels that can be produced with such waste products is limited by their availability, and what is often considered a waste product often has at least some economic value. Consider, for example, the case of making ethanol from molasses, a by-product of sugar production. While molasses has a much lower value than processed sugar, it does have alternative uses such as livestock feed. Thus while the food security concerns associated with producing ethanol from molasses may be less than the concerns associated with producing ethanol from grain or biodiesel from vegetable oil, there may still be some food security implications to consider.

The lesson from these various examples is that the food security implications of biofuel production are often quite complicated. Most current biofuel production technologies will have at least some impact on food availability and prices. Benefits of biofuel production must be weighed carefully against possible ill consequences.

5. Conclusion

This chapter discussed the three pillars of sustainability–social, economic, and ecological—with respect to biomass development. Economic viability and sustainability can be measured by profitability and efficiency. In the context of profitability, production cost should be less than the price of biofuel, and in terms of efficiency, the maximum amount of yield should be achieved within the given resources. Also, in the commercial context, the price of biofuel should be lower compared to the price of other equivalent oils to maintain competitiveness in the biofuel market. The 2007–08 food crisis heightened the dilemma between food security and biofuel development (availability and access to food). It has certainly increased agricultural commodity prices (sharing the same productive resources). However, there are other factors that may step in to moderate the price increase that may happen due to increased demand from biofuels. Also, the crops that are used to produce biofuels are less domestic and more globally traded commodities, so it would influence the market price by moderating it. In the industrial context, biofuel production has the potential to affect and improve market mechanisms, for example, physical infrastructure, which in turn can moderate prices.

Given that the development of bioenergy is supposed to increase energy security and to understand to what extent biomass could replace fossil fuels, the concept of fossil energy came into being. By definition, it is the ratio of the output of renewable energy and the input of fossil energy required for the production of biofuel.

The environmental sustainability of biofuels plays an important role in reducing GHG emissions, water management, soil erosion, biodiversity, and energy balance. Mitigation of GHG emissions is considered to be one of the most significant positive environmental impacts. Moreover, there is a concern that biofuel GHG reduction is not markedly efficient if the conversion happens from forest and grassland into agricultural land. On the other hand, biomass deployment from degraded land may have a positive effect.

This chapter discussed the economic, environmental, and social dimensions of sustainability in the context of biomass development. Economic viability of biomass can be assessed in terms of profitability (the price of biofuel exceeds the production cost) and efficiency (the maximum amount of yield is obtained with a given quantity of resources). The economic competitiveness of biofuels calls for their price to remain below the price of oil equivalents. The food crisis of 2007–08 heightened the debate on biofuels development and food security (availability of and access to food). Increased demand from biofuels has an impact on agricultural commodity prices. However, several factors could moderate any price increases that may occur. Because most crops currently used for biofuels are globally traded commodities, market competition will moderate prices. The biofuels industry has the potential to create and improve market mechanisms such as physical infrastructure, which can moderate prices.

Given that one important motivation for bioenergy development is to increase energy security and the need to understand the extent to which biomass is qualified to replace fossil fuels, the notion of fossil energy balance was introduced: the ratio between renewable energy output and fossil energy input needed to produce the biofuel.

The environmental sustainability of biofuels is described in terms of their implications for energy balance, GHG emissions, biodiversity, water, and soil. Reduction of GHG emissions is considered to be the most significant environmental impact. However, biofuels' GHG reduction potential suffers markedly from any conversion of grasslands and forests into agricultural land.

Biodiversity is recognized as an important factor. Biomass production for bioenergy can have both positive and negative impacts on biodiversity. When degraded land is used and if GHG emissions are reduced, the diversity of species might be enhanced; on the other hand, large monocultures of energy crops can cause habitat loss, the expansion of invasive species, and contamination from fertilizers and herbicides, with concomitant erosion of biodiversity.

References

Abbott, P., Hurt, C., Tyner, W., 2008. What's Driving Food Prices? Farm Foundation. https://www1.eere.energy.gov/bioenergy/pdfs/farm_foundation_whats_driving_food_prices.pdf.

Alberti, M., 1996. Measuring Urban Sustainability. Environ. Impact Assess. Rev. 16, 381–424.

Broch, A., Hoekman, S.K., Unnach, S., 2013. A review of variability in indirect land use change assessment and modeling in biofuel policy. Environ. Sci. Policy 29, 147–157.

Brundtland Report, 1987. Our Common Future. WCED 1987. p. 43.

Callaghan, G.E., Colton, J., 2008. Building sustainable & resilient communities: a balancing of community capital. Environ. Dev. Sustain. 10, 931–942.

Columella, L.J.M., 1948. Res rustica. Lucius Junis Moderatus Columella on agriculture. Vol. 1. Harvard University Press, Cambridge, MA (Books I-IV). English translation by Harrison Boyd Ash. 1948 reprint.

Debnath, D., Whistance, J., Thompson, W., 2017. The causes of two-way US-Brazil ethanol trade and the consequences for greenhouse gas emission. Energy 141, 2045–2053.

de Gorter, H., Drabik, D., Just, D., 2013. How biofuels policies affect the level of grains and oilseed prices: theory, models and evidence. Global Food Secur. 2 (2), 82–88.

EPA, 2018. Biofuels and the Environment: The Second Triennial Report to Congress. https://www.researchgate.net/publication/326548741_Biofuels_and_the_Environment_The_Second_Triennial_Report_to_Congress.

FAO (Food and Agricultural Organization of the United Nations). 2011, The Global Bioenergy Partnership Sustainability Indicators for Bioenergy, first ed. FAO.

Hochman, G., Rajagopal, D., Timilsina, G., Zilberman, D., 2014. Quantifying the causes of the global food commodity price crisis. Biomass Bioenergy 68, 106–114.

Jevons, S.W., 1866. The Coal Question, An Inquiry Concerning the Progress of the Nation, and the Probable Exhaustion of Our Coal-Mines, second ed. Macmillan & Co.

Morelli, J., 2011. Environmental sustainability: A definition for environmental professionals. J. Environ. Sustain. 1(1).

Menalled, F., Bass, T., Buschena, D., Cash, D., Malone, M., Maxwell, B., Kent, M.V., Miller, P., Soto, R., Weaver, D., 2008. An Introduction to the Principles and Practices of Sustainable Farming. Montana State University, Extension.

Overmars, K., et al., 2015. Estimates of Indirect Land Use Change From Biofuels Based on Historical Data. Publications Office https://ec.europa.eu/jrc/en/publication/eur-scientific-and-technical-research-reports/estimates-indirect-land-use-change-biofuels-based-historical-data.

Pisani, A.D.J., 2006. Sustainable development-historical roots of the concept. Environ. Sci. 3 (2), 83–96.

Robinson, J., 2004. Squaring the circle? Some thoughts on the idea of sustainable development. Ecol. Econ. 48, 369–384.

Searchinger, T., Heimlich, R., Houghton, R.A., Dong, F., Elobeid, A., Fabiosa, J., Yu, T.-H., et al., 2008. Use of US croplands for biofuels increases greenhouse gases through emissions from land-use change. Science 319 (5867), 1238–1240.

Taheripour, F., Tyner, W.E., 2013. Biofuels and land use change: applying recent evidence to model estimates. Appl. Sci. 3, 14s–38.

U.S. Department of Agriculture, 2018. World Agricultural supply and Demand Estimates. Various Reports. Washington D.C.

U.S. Energy Information Administration, 2018. Monthly Energy Review. Various Reports. Washington D.C.

Westhoff, P., 2010. The Economics of Food: How Feeding and Fueling the Planet Affects Food Prices. FT Press, New Jersey.

Walsh, R.P., 2011. Creating a "values" chain for sustainable development in developing nations: where Maslow meets Porter. Environ. Dev. Sustain. 13, 789.

Further reading

Colantonio, A., 2009. Measuring Social Sustainability: Best Practice from Urban Renewal in the EU. Oxford Institute for Sustainable Development (OISD) – International Land Markets Group.

CHAPTER 13

Advanced biofuels: Supply chain management

Deepayan Debnath
Food and Agricultural Policy Research Institute, University of Missouri, Columbia, MO, United States

Contents

1. Introduction

With an increase in the prices of fossil fuels and concern over environmental degradation there has been increased interest in finding alternative sources of energy. Part of this interest has focused on renewable bioenergy, which is expected to have fewer negative environmental consequences than hydrocarbon fuels. However, because of the bulky nature of the feedstocks used to derive the renewable energy, more specifically cellulosic feedstock, it is very important to understand the supply chain of advanced biofuel feedstocks.

The standard paradigm for evaluating the economics of cellulosic biofuels has largely followed the pattern used to evaluate grain ethanol. However, producing, harvesting, storing, and delivering cellulosic biomass from a dedicated perennial energy crop such as switchgrass and miscanthus and converting it to biofuel is fundamentally different to producing and marketing grains, and producing biofuel from those grains. The infrastructure for maize grain was well developed prior to implementation of public policies designed to increase the production of fuel ethanol. Initially, the quantity of grains

Biofuels, Bioenergy and Food Security
https://doi.org/10.1016/B978-0-12-803954-0.00013-9
231

(maize and wheat) used by ethanol plants relative to total global grain production was relatively small. In bumper crop years, as well as short crop years, ethanol biorefinery managers could bid for grains from alternative uses. When the proportion of the crops that they required to meet their capacity requirements was relatively small, except for having to pay a higher price for feedstock, they did not have to be overly concerned about the risk of idling the biorefinery due to a short crop. However, a similar feedstock production and delivery infrastructure does not exist for potential cellulosic biomass biorefineries.

Procurement of biomass from dedicated perennial energy crops such as switchgrass and miscanthus will be fundamentally different. A similar feedstock production and delivery infrastructure does not exist for potential cellulosic biorefineries designed to use advanced biofuel feedstocks, including switchgrass and miscanthus biomass. Prior to investing in a cellulosic biorefinery, prudent investors would expect assurance that a flow of biomass that meets the quality standards of the facility will be available at a cost that provides a high probability of a good investment. The first biorefinery in a region will not be able to rely on spot markets to procure feedstock since spot markets do not exist. One potential strategy would be for the biorefinery to engage in long-term contracts with growers or land owners. Since major energy crops are perennial and the annual harvestable biomass yield is uncertain, land owners are much more likely to be willing to contract for land areas to seed energy crops than for delivering a prespecified quantity every year. A single biorefinery would require that several thousand hectares within a reasonable distance of the proposed biorefinery be changed from existing use and seeded for energy crops. Another alternative would be for the biorefinery to engage in long-term land leases with land owners and for the biorefinery to manage the harvest and delivery of feedstock. In either case, biorefinery management would be required to identify and contract for a sufficient quantity of land or production in the vicinity of the biorefinery to provide for the planned feedstock needs.

If the annual feedstock requirements of the biorefinery and annual energy crops yield were known with certainty, it would be straightforward to determine the number of hectares to lease. However, advanced biomass, mainly switchgrass and miscanthus, yields vary from year to year. In years with unfavorable biomass production weather, yields in the feedstock supply shed of the biorefinery may be low, and if too few hectares are leased, production from the leased hectares may be insufficient to meet the needs of the biorefinery. Since advanced biomass cannot be anticipated to be available from spot markets, the biorefinery may be forced to shut down for a period of time. Each idle day for lack of feedstock will have economic consequences. The net opportunity cost of a forced idle day (downtime cost) will depend on the lost revenue as well as on fixed production costs that cannot be avoided.

On the other hand, the biorefinery could plan to maintain a storage reserve from which biomass may be retrieved in low production years. However, there are a number of issues associated with maintaining the quality of switchgrass biomass for an extended period of time. The cost and efficiency of conversion for systems such as enzymatic hydrolysis depend on the characteristics of the feedstock. Maintenance of feedstock quality over an extended period of time would require expensive storage facilities, added handling costs, and carry a substantial risk of loss from fire. Additionally, managing consecutive bad weather years could be challenging and require a large investment.

In either case, a system will be necessary to identify and contract for a sufficient quantity of land or production in the vicinity of the biorefinery to provide for planned feedstock needs. Selection of a biorefinery location and identification of land to lease for the production of feedstock will be critical decisions in determining the economic success of an advanced biomass processing facility.

2. Risk and variability

Biomass yield from perennial grasses varies considerably across regions and across years. Therefore it is important to trace the potential consequences of biomass spatial and temporal yield variability on the optimal quantity, quality, and location of land to lease, and on the cost of delivering a flow of biomass from the perspective of the biorefinery. If land leasing decisions are based on average yields and if yields are highly correlated across the land leased, in bad weather years, yields from the leased land may be insufficient to meet the needs of the biorefinery and the biorefinery would be forced to shut down for a period of time, which would entail economic consequences.

In good weather years, yields may be greater than expected and more biomass may be produced than can be processed. Excess biomass may be mowed and left in the field to build organic matter. However, it will cost the producer more and eventually increase the price of feedstocks. The cost of harvesting and storing the excess could be avoided. Alternatively, the excess may be harvested and stored for use in the future. This strategy will incur additional harvest and storage costs and losses. The optimal strategy will depend in part on the expected costs of harvest, storage, and expected storage losses. If yields across fields within the potential biorefinery supply shed are not highly correlated, a biorefinery might attempt to reduce overall year-to-year variability in feedstock production by strategically selecting and leasing a portfolio of land. Therefore a land leasing strategy is required that considers both year-to-year biomass yield variability and within-year yield variability across the supply shed of a potential cellulosic biomass biorefinery.

3. Supply chain model

In this section, several innovative supply chain optimization models are discussed for a given biorefinery that is capable to determine the optimal quantity, quality, and location of land to lease considering yield variability. First, a traditional model designed to minimize the cost to meet the expected annual biorefinery requirements based on average switchgrass yields (model 1) is described. Second, an enhanced model designed to identify the optimal quantity, quality, and location of land to contract to ensure that a sufficient quantity of feedstock can be delivered even in the worst expected production state of nature is presented (model 2). Third, a model is developed that enables determination of the tradeoff between the opportunity costs of closing the biorefinery as a result of insufficient feedstock and the quantity of contracted land considering the yield variability (model 3). This model is innovative because it enables the determination of the optimal quantity, quality, and location of land to lease (L) considering yield variability for a given level of biorefinery downtime (D) opportunity cost for a proposed biorefinery. The estimate of the opportunity cost of downtime may be parameterized to trace out the DL frontier; that is, the downtime cost–land to lease frontier. A fourth model that enables year-to-year storage along with downtime is also formulated. The models encompass both spatial (across land class and across zones within each year) and temporal (across years) biomass yield variability. Findings from each of the models may be compared across models to determine the circumstances for which energy crop production cost varies over each production activity including raking, baling, transportation, and storage.

3.1 Model 1: Identifying land to contract based on average switchgrass yields

In model 1, the objective function is optimized subject to available land, average energy crop (switchgrass or miscanthus) biomass yields, and the annual biorefinery requirements. The objective is:

$$\min_{XL, XB, XT} C = \sum_{i}^{I} \sum_{j}^{J} \lambda_{ij} XL_{ij} + \beta \sum_{i}^{I} \sum_{j}^{J} XB_{ij} + \sum_{i}^{I} \sum_{j}^{J} \tau_{ij} XT_{ij} \qquad (13.1)$$

where C is the average costs per year of contracting land, producing, harvesting, and transporting a fixed quantity of biomass to a predetermined location to meet the biorefinery demand, λ_{ij} is the annual production cost for amortized establishment costs, land rent, fertilizer, mowing, and raking costs in zone i for land class j, XL_{ij} is the quantity (ha) of land class j leased for energy crop production in zone (district/county) i, β is the per unit cost of baling and stacking energy crop biomass, XB_{ij} is the quantity (Mg) of switchgrass or miscanthus baled and stacked in zone i from land class j, τ_{ij} is the per Mg cost of loading and transporting switchgrass biomass from zone i and land class j to the biorefinery, and XT_{ij} is the quantity (Mg) of energy crop biomass transported from zone i and land class j to the biorefinery.

Eq. (13.1) is minimized subject to the following set of constraints.

$$XL_{ij} \leq \eta_{ij} \quad \forall i,j \tag{13.2}$$

Eq. (13.2) imposes land constraints where η_{ij} is the quantity (ha) of land class j in zone i available for contracting from current use for conversion to bioenergy crop production.

$$XB_{ij} \leq \overline{\alpha}_{ij} XL_{ij} \quad \forall i,j \tag{13.3}$$

Eq. (13.3) balances the quantity baled with the quantity produced; $\overline{\alpha}_{ij}$ is the mean switchgrass or miscanthus biomass yield (Mg/ha/year) in zone i from land class j over T years.

$$XT_{ij} \leq XB_{ij} \quad \forall i,j \tag{13.4}$$

Eq. (13.4) balances the quantity transported with the quantity baled.

$$\sum_i \sum_j XT_{ij} \geq \delta \tag{13.5}$$

Eq. (13.5) imposes the requirement that the quantity transported to the biorefinery fulfills the annual switchgrass or miscanthus feedstock requirement, denoted by the scalar δ (Mg/year).

$$XL_{ij}, XB_{ij}, XT_{ij} \geq 0 \tag{13.6}$$

Eq. (13.6) is included to restrict the choice variables to be nonnegative.

Model 1 is designed to select the optimal (least-cost) location (district/county), quality (land class), and quantity of land on which to establish energy crops to fulfill the biorefinery's annual requirement under the restrictive assumption that the average yield will be obtained on each land class in each zone (district/county) and each year. If yields are normally distributed and perfectly correlated across the land classes and districts or counties in the supply shed, in some years the land identified by the model for contracting would produce more than the biorefinery could process. Alternatively, in other years, production from the contracted land would be insufficient to meet the annual needs and the biorefinery would have insufficient feedstock to operate at full capacity throughout the processing season (year).

Rather than contracting land based on average yields, the biorefinery could choose to contract sufficient land so that even in the expected worst case production situation, adequate feedstock would be produced to enable the biorefinery to operate at full capacity in each year. While the true worst-case production situation cannot be known, historical data may be used to generate biomass yield distributions. To the extent that the generated yield distributions capture future yield variability and are available for each land class and zone in the supply shed, a model may be formulated to identify which land to contract to ensure that the biorefinery could operate at full capacity even in poor feedstock production years. Model 2 includes constraints that require biorefinery requirements to be fulfilled for each possible state of nature based on available historical yield distributions and to determine which land should be contracted to minimize the cost to deliver feedstock.

3.2 Model 2: Identifying land to contract to fulfill requirements in all states of nature

Model 2 recognizes that yield from each land class in each zone will differ in each state of nature but that for each state of nature the fixed demands of the biorefinery must be met. The model is designed to identify the location (district/county), quality (class), and quantity of land to contract. Every parcel that is contracted must be mowed in every year. However, for each year, depending on production, a unique combination of the contracted hectares may be optimally raked and baled. Thus transportation flows may differ every year. The model is designed to compute the minimum expected cost over the T states of nature. The T states of nature are assumed to represent the complete yield distribution for each land class and each zone. The objective function follows:

$$\min_{XL, XR, XB, XT} EC = \sum_i^I \sum_j^J \gamma_{ij} XL_{ij} + \rho \left(\sum_t^T \sum_i^I \sum_j^J XR_{tij} \right)/T + \beta \left(\sum_t^T \sum_i^I \sum_j^J XB_{tij} \right)/T + \left(\sum_t^T \sum_i^I \sum_j^J \tau_{ij} XT_{tij} \right)/T$$

(13.7)

where EC is the expected costs per year of renting land, producing, harvesting, and transporting biomass to a predetermined location to meet the biorefinery demand in each of the T states of nature, γ_{ij} is the average annual production cost, including amortized establishment costs, land rent, fertilizer, and mowing costs in zone i for land class j, ρ is the cost (per ha) to rake mowed biomass into a windrow for baling ($\gamma_{ij} = \lambda_{ij} - \rho$), and XR_{tij} is the quantity (ha) of land class j raked in year t in zone i. A choice variable for raking is introduced, since by assumption all of the land that is contracted must be mowed once per year. However, raking is required only for biomass that must be baled to fulfill the annual requirements. In years when production exceeds requirements, the excess production is mowed and left in the field to decompose. Since the cost of raking depends on the area (ha) of land raked while the cost to bale is a function of yield (Mg), raking and baling activities are considered separately. T is the number of states of nature (years) for which historical yields are available, XB_{tij} is the quantity (Mg) of switchgrass baled and stacked in year t in zone i from land class j, XT_{tij} is the quantity (Mg) of switchgrass biomass transported from zone i and land class j to the biorefinery in year t.

Eq. (13.7) is minimized subject to the following constraints:

$$XL_{ij} \leq \eta_{ij} \quad \forall i, j$$

(13.8)

$$XR_{tij} \leq XL_{ij} \quad \forall i, j$$

(13.9)

Eq. (13.9) restricts the area (ha) raked to be less than or equal to the area under contract. In a given year, only the quantity of biomass required to fulfill the needs of the biorefinery will optimally be raked. The area raked may differ in each state of nature.

$$XB_{tij} \leq \alpha_{tij} XR_{tij} \quad \forall t, i, j$$

(13.10)

Eq. (13.10) restricts the quantity to be baled in year t from land class j in zone i to be no more than the yield (α_{tij}) in year t from land class j in zone i times the area raked.

$$XT_{tij} \leq XB_{tij} \quad \forall t, i, j \tag{13.11}$$

Eq. (13.11) balances the quantity baled with the quantity transported to the biorefinery from land class j in zone i for each year t.

$$\sum_i \sum_j XT_{tij} \geq \delta \quad \forall t \tag{13.12}$$

Eq. (13.12) imposes the constraint that the quantity transported to the biorefinery fulfills the annual switchgrass feedstock requirement in every year.

$$XL_{ij}, XR_{tij}, XB_{tij}, XT_{tij} \geq 0 \tag{13.13}$$

Eq. (13.13) restricts the choice variables to be nonnegative.

Model 2 imposes the requirement that a sufficient quantity of land be contracted so that even in the most unfavorable weather production situation, biomass will be produced to enable the biorefinery to operate at full capacity for the entire year. This requires contracting and paying for more land than would be required in all except the most extreme bad weather years. An alternative would be to contract for less land and in years when production was less than sufficient, to idle the plant when feedstock is not available. However, this strategy would reduce the quantity and cost of land that is contracted for bioenergy production.

3.3 Model 3: Identifying the D (downtime cost)-L (land to lease) frontier

Model 3 is designed to determine the economic tradeoffs that result from contracting for less land resulting in forced idle days in certain years. The objective function follows:

$$\min_{XL, XR, XB, XT, XS} EAC = \sum_i^I \sum_j^J \gamma_{ij} XL_{ij} + \rho \left(\sum_t^T \sum_i^I \sum_j^J XR_{tij} \right) / T$$
$$+ \beta \left(\sum_t^T \sum_i^I \sum_j^J XB_{tij} \right) / T + \left(\sum_t^T \sum_i^I \sum_j^J \tau_{ij} XT_{tij} \right) / T + \vartheta \left(\sum_t^T XS_t \right) / T \tag{13.14}$$

where EAC includes the expected annual cost of not meeting the biorefinery demand as well as the average costs per year of contracting land, producing, harvesting, and transporting biomass to the biorefinery, ϑ is the per Mg penalty (opportunity cost) of not delivering sufficient feedstock to meet biorefinery requirements, and XS_t is the quantity (Mg) of switchgrass less than the biorefinery capacity that is not available for processing in year t (shortage). Other variables are as previously defined.

The model includes the following constraints:

$$XL_{ij} \leq \eta_{ij} \quad \forall i, j \tag{13.15}$$

$$XR_{tij} \leq XL_{ij} \quad \forall i, j \tag{13.16}$$

$$XB_{tij} \leq \alpha_{tij} XR_{tij} \quad \forall t, i, j \tag{13.17}$$

$$XT_{tij} \leq XB_{tij} \quad \forall t, i, j \tag{13.18}$$

Eqs. (13.15)–(13.18) are as previously defined.

$$\sum_i \sum_j XT_{tij} \geq \delta - XS_t \quad \forall t \tag{13.19}$$

For a given year, Eq. (13.19) permits the annual biorefinery capacity to be relaxed by the quantity XS_t (Mg).

$$\sum_t XS_t \leq \varphi \quad A = \pi r^2 \tag{13.20}$$

The total shortage quantity (Mg) across all years that may be permitted is constrained to a level, φ, as shown in Eq. (13.20). If the value of φ is set equal to zero, the model will identify sufficient land to be contracted to deliver δ Mg of feedstock in every year and model 3 will provide the identical solution as model 2. To allow for average feedstock shortages that would result in 1 shutdown day per year, the value of φ may be set equal to one times the daily capacity times the number of states of nature that make up the empirical switchgrass yield distributions. The value of φ may be changed in combination with the value of v in Eq. (13.14) to trace the tradeoff between land area contracted and the cost to idle the biorefinery.

$$\sum_i \sum_j \alpha_{tij} XR_{tij} + XS_t - XE_t = \delta \quad \forall t \tag{13.21}$$

Eq. (13.21) balances the feedstocks in each year. In years when production exceeds biorefinery requirements, XE_t tracks the quantity (Mg) of excess production and will be greater than zero. In years when production is short of biorefinery requirements, XS_t tracks the shortage quantity (Mg) and will be greater than zero.

$$XL_{ij}, XR_{tij}, XB_{tij}, XT_{tij}, XS_t, XE_t \geq 0 \tag{13.22}$$

Eq. (13.22) restricts the choice variables to be nonnegative.

3.4 Model 4: Incorporating interyear storage

The model is designed to determine the economic tradeoffs that result from leasing less land resulting in forced idle days in certain years in the presence of storage capacity (XC_{tij}). The objective function follows:

$$\min_{XL, XR, XB, XT, XS, XC} ESC = \sum_i^I \sum_j^J \gamma_{ij} XL_{ij} + \rho \left(\sum_t^T \sum_i^I \sum_j^J XR_{tij} \right)/T + \beta \left(\sum_t^T \sum_i^I \sum_j^J XB_{tij} \right)/T$$

$$+ \sum_t^T \sum_i^I \sum_j^J \tau_{ij} XT_{tij})/T + \vartheta \left(\sum_t^T XS_t \right)/T + \sigma \left(\sum_t^T \sum_i^I \sum_j^J XC_{tij} \right)/T \tag{13.23}$$

where *ESC* includes the expected annual opportunity cost of not meeting the biorefinery demand as well as the average costs per year of leasing land, producing, harvesting, storing, and transporting biomass to the biorefinery, XC_{tij} is the quantity (Mg) of switchgrass or miscanthus storaged in year t in zone i produced on land class j, and σ is the cost to storing 1 Mg of switchgrass for 1 year. Other variables are as previously defined. Eq. (13.23) is optimized subject to Eqs. (13.15)–(13.17), (13.19), (13.24)–(13.26). Eq. (13.24) restricts the quantity to be stored in year t from land class j in zone i to be no more than the quantity to be baled in year t from land class j in zone i.

$$XC_{tij} \leq XB_{tij} \quad \forall t, i, j \tag{13.24}$$

Eq. (13.25) balances the quantity baled in period t plus the quantity available in storage from prior periods after losing (ϕ) percentage of biomass in storage per year minus quantity transported to the biorefinery in period t with the quantity stored for use in period $t+1$.

$$XC_{t+1ij} = XB_{tij} + (1-\varphi)XC_{tij} - XT_{tij} \quad \forall t, i, j \tag{13.25}$$

Eq. (13.26) includes the nonnegativity constraints.

$$XL_{ij}, XR_{tij}, XB_{tij}, XT_{tij}, XS_t, XC_t \geq 0 \tag{13.26}$$

4. Estimating energy crop biomass yield distributions

Energy crop (switchgrass or miscanthus) historical yields are not available. However, a proxy for historical yields may be simulated by a calibrated and validated biophysical plant growth simulation model if field trial yield data and historical weather data are available. The Environmental Policy Integrated Climate (EPIC) model (Williams, Jones, and Dyke, 1984; Mondzozo et al., 2011) can be calibrated for switchgrass or miscanthus yields and validated against biomass field trial yield data obtained from any specific geographical location. Calibration and validation of biomass yields were performed for each of the different land classes (soil types). Soil-related information for these land types, including bulk density, water, clay, sand and silt content, organic carbon concentration, calcium carbonate content, saturated conductivity, and cation exchange capacity can be obtained from the SSURGO land database in the US context, while the Food and Agricultural Organization maintain the world soil database. The field trial data used to validate the calibrated results might be obtained from annual and biannual harvests. Therefore while setting the biomass yield simulation in EPIC, special attention on the harvest rate is important to calibrate it with the corresponding field trial experiments. Daily weather data of solar radiation, maximum temperature, minimum temperature, relative humidity, wind velocity, and precipitation for each of the zones for each year for which data were available were obtained from the national weather data archives.

Once the EPIC model is calibrated it is ready to use to simulate switchgrass biomass yields (α_{tij}) for the states of nature (historical years or climate change scenarios) for different types of land for each of the zones. Soil classification within each land class with the most hectares in the counties or districts can be used in the EPIC simulation to represent the specific land class. In other words, it can be assumed that all soil types in the zone within a particular land class produced identical yields in each simulated year. Thus weather, soil, and management information proxies for empirical yield distributions for each state of nature were produced for each zone and each land class.

5. EPIC model calibration and validation: An example of switchgrass in Oklahoma, USA

Productivity, growth, longevity, and adaptation traits of switchgrass primarily depend on the geographical location of their origin. Based on the latitude and longitude of origin, switchgrass is broadly classified into two ecotypes: upland and lowland. Lowland ecotype varieties are more compatible with the southern latitude or southern part of the United States due to their ability to adapt to the longer growing season and warmer climatic conditions. Upland varieties are widely adapted in the northern part of the United States due to their greater potential to survive in the colder conditions of northern latitudes (Casler et al., 2007).

Lowland ecotype switchgrass yields were calibrated using EPIC v.0509 and validated against switchgrass biomass field trial yield data obtained from three locations: Chickasha, Haskell, and Stillwater. Calibration and validation of switchgrass yields were performed for each of the three different soil types: McLain silt loam, Taloka silt loam, and Kirkland silt loam on which the field experiments at Chickasha, Haskell, and Stillwater were conducted. Soil-related information for these land types, including bulk density, water, sand and silt content, organic carbon concentration, calcium carbonate content, saturated conductivity, and cation exchange capacity were obtained from the SSURGO land database. The field trial data used to validate the calibrated results were obtained from a single annual harvest. Single harvest management practices starting from the second year, as conducted in the field experiments, were used to calibrate the EPIC model. Daily weather data of maximum temperature, minimum temperature, precipitation, solar radiation, relative humidity, and wind speed for each location were obtained from MESONET (2011) and NOAA (2011).

Calibration required adjustments to the EPIC crop parameter (CROPCOM crop file in EPIC v.0509). Timing of leaf decline (DLAI, EPIC v.0509 crop parameter acronyms), and maximum leaf area index (DMLA) were adjusted to 0.75 and 6, respectively. Leaf area decline after anthesis (RLAD), rate of decline in biomass energy after anthesis (RBMD), and plant maturity (RWPC2) were adjusted to 0.1, 0.1, and 0.3, respectively (Thomson et al., 2009).

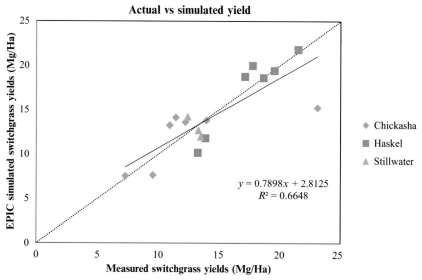

Fig. 13.1 Comparison between EPIC simulated and measured lowland ecotype switchgrass yields for the three locations of Oklahoma over the year 1994 through 2000 and 2003–05. *EPIC*, environmental policy integrated climate.

The Chickasha and Haskell field trials included three lowland ecotype switchgrass cultivars: Alamo, Kanlow, and PMT 279. Measured yields of these cultivars were averaged and compared to the simulated yields. The Stillwater trials included only Alamo. EPIC-simulated switchgrass yields were compared against the actual switchgrass yields (Fig. 13.1). The simulated yields explained 67% of the variation in the measured yields for the 10 years (1994–2000 and 2003–05). However, the model did not closely predict the Chickasha yields recorded for 1995. When the 1995 Chickasha observation was dropped, the R^2 increased to 0.84. By this measure, the model was assumed to have successfully captured the switchgrass biomass yield response and yield variation and was assumed to be calibrated.

6. Land availability and values

Bioenergy crop cultivation on agricultural land is hugely criticized as the debate over "food versus fuel" escalates. Scientists started proposing the idea of growing energy crops on "marginal land." However, the definition of "marginal" land is vague. Marginal lands can be defined as soils that have physical and chemical problems or are uncultivated or adversely affected by climatic conditions. This may be why it is frequently used in discussions of dedicated energy crop production even though there was no consistent definition (Lewis and Kelly, 2014). After reviewing 51 studies published between 2008 and 2012 that included the term "marginal land(s)" and/or "marginal soil(s)," Richards et al. (2014) identified that only half of them provided a clear definition of "marginal." They found that often in the papers the term "marginal" was subjectively labeled for less-than-ideal lands. In the most extreme case, the word "marginal" only appeared in the title of the paper. This clearly suggested ambiguity in terms of the classification of marginal lands. Therefore to avoid this ambiguity over the use of "marginal land" for the production of bioenergy crops, one approach is to constrain the total available agricultural land to produce energy crops to 10%. In the case of the aforementioned models, the quantity of land (XL_{ij}) assumed to be available for contracting by the biorefinery could be set equal to 10% of the total quantity of each land class in each zone.

The value of any agricultural land devoted to the production of energy crops is defined as the opportunity cost of producing the existing crop on that particular zone (county or district) of land. In the case of Oklahoma in the United States, the estimated land cost for each land class for each county is obtained by extrapolating the US Department of Agriculture (USDA) cropland rental values using Eq. (13.27). Land rent cost for each land class in each county is estimated to be:

$$\omega_{ij} = \mu_{ij}\left(\frac{\sum_{j}X_{ij}\chi_i}{\sum_{j}\mu_{ij}X_{ij}}\right) \tag{13.27}$$

where ω_{ij} is the rental cost of a hectare of land in county i and land class j, μ_{ij} is the potential wheat yield in county i and land class j as reported in the SSURGO database, X_{ij} is the available hectares of land in county i and land class j, and χ_i is the USDA reported cropland rental rate for county i in the state on Oklahoma in the United States.

7. Transportation costs

Transportation costs (τ_{ij}) depend on the distance traveled. Therefore it is calculated based on the following linear equation:

$$\tau_{ij} = a + bD_{ij} \tag{13.28}$$

where τ_{ij} is the estimated costs ($/Mg) for loading and transporting 1 Mg of switchgrass dry matter from land class j of zone i to the biorefinery and D_{ij} is the one-way distance (km) between the centroid of land class j of zone i and the biorefinery. The centroid of each land class in each zone is determined and the nearest town to the corresponding land class centroid can be obtained via a geographical information system. Road distance between the town and the biorefinery was obtained from Google Maps (maps.google.com).

8. Cost of idling the biorefinery

The net amount of revenue lost due to not delivering 1 Mg of feedstock that could be processed will depend on the lost revenue as well as on the variable production costs that can be avoided. If it is assumed that none of the biorefinery operating costs can be avoided, lost revenue from not producing biofuel will depend on the net market price of biofuel. For a conversion rate of 375 L/Mg, and a wholesale biofuel price of $0.55/L (which is $0.715 in gasoline energy equivalent level), the penalty charges for not processing feedstock derived from the lost revenue would be $206/Mg. This provides an estimate of the value of ϑ in Eq. (13.14). Since some of the operating costs may be avoided if the biorefinery is idled, this can be interpreted as an upper bound estimate of the penalty for not processing 1 Mg of feedstock.

9. An example of a hypothetical biorefinery in Oklahoma

A biorefinery was assumed to be located near Okemah in Okfuskee County, Oklahoma, USA (Fig. 13.2). Based on estimates provided by Kazi et al. (2010) and Wright et al. (2010), the biorefinery daily requirement of feedstock was set equal to 2000 Mg. Assuming downtime required for maintenance and allowing for 350 days of operation per year, the annual feedstock requirement (δ) was set at 700,000 Mg.

When the model is solved for the particular study area, in this case Oklahoma, USA, as the yield varies across time and space, the location of fields to bale and biomass to transport varies from year to year. Fig. 13.1 shows that in the case of this hypothetical biorefinery in Oklahoma the model that considers the temporal switchgrass yield variability results in leasing more land to grow switchgrass and at a further distance over the model solved with average yield. Fig. 13.3 shows that harvesting costs (including mowing, raking, and baling activities) and transportation cost are the key components in the supply chain of the advanced biorefinery mainly due to the bulky nature of cellulosic feedstocks.

10. Conclusion

Much has been said about the anticipated cost reductions expected to be achievable with the *n*th refinery relative to the first cellulosic biorefinery with a specific technology (Kazi et al., 2010; Wright et al., 2010). However, these analyses have failed to consider that unlike cookie cutter maize-based ethanol plants that can procure a flow of feedstock by simply offering a price premium for maize grain relative to the local elevator, each cellulosic biorefinery will need to pay careful attention to feedstock procurement. Transportation costs can be expected to limit the procurement region for a biorefinery. Biomass yield from perennial grasses varies considerably across regions and across years (Sala et al., 1988). In many cases, failure to consider spatial and temporal yield variability could be quite costly. If the initial biorefineries are located in regions with less biomass yield variability, then the *n*th biorefinery may be located in a region with more yield variability and greater feedstock cost. The modeling system presented herein is generalizable to any regions of the world and is not constrained with any type of specific perennial dedicated energy crops such as switchgrass and miscanthus. These models can even be used for short

Land leased under average switchgrass yield (model 1)

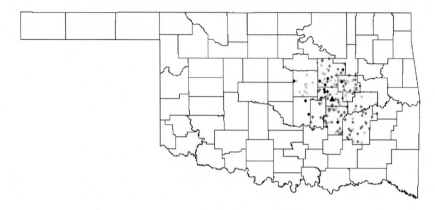

Land leased to ensure sufficient feedstock in each production year (model 2)

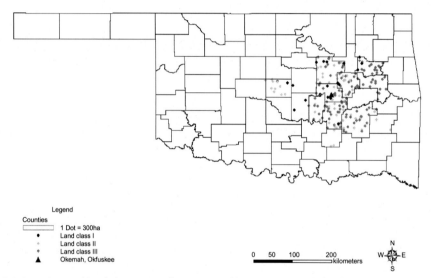

Legend

Counties

1 Dot = 300ha
• Land class I
· Land class II
• Land class III
▲ Okemah, Okfuskee

0   50   100   200 kilometers

Fig. 13.2 Location and land class optimally contracted by county. (Symbols randomly assigned within counties).

rotations of woody crop operations too. The value of strategically selecting land to contract will vary across regions and across species.

Development of feedstock production could be expected to develop simultaneously with biorefinery construction. A biorefinery designed to process switchgrass or miscanthus feedstock could engage in long-term contracts designed to fulfill its feedstock needs. The models presented in this chapter account for the spatial and temporal variability of switchgrass biomass yield, and address one of the most complex issues that potential biorefinery investors are facing.

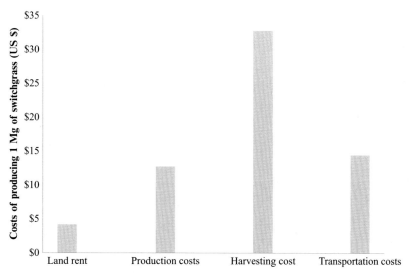

Fig. 13.3 Land rent, production costs, harvest costs, and transportation costs of 1 Mg of switchgrass.

Models based on average production do not account for the opportunity cost incurred by the biorefinery for not producing and selling products in those years when biomass production from contracted land is not sufficient to enable the biorefinery to operate at production capacity. A fully vertically integrated system may reduce the variability of feedstock cost but cannot eliminate it.

Companies may be reluctant to contract for sufficient quantities of land to provide for feedstock needs. Public opinion may not support conversion of 50,000 ha in a region from existing use to the production of a dedicated energy crop. The probability of successful rent-seeking behavior on behalf of the biorefinery may be reduced if it becomes clear that production of perennial grass-based feedstock is more akin to harvesting non-agricultural lands than farming conventional annual crops. Public officials may place impediments limiting the ability of biorefineries to lease large tracts of land. However, ambiguities as to what determines feedstock quality and how to provide a flow of feedstock throughout the year are likely to be resolved much more quickly if the annual payment to the land owner is set. Leased land would enable the biorefinery to manage both feedstock quality and harvest and transportation to optimize the field for a biofuel system.

If an economically viable system for converting biomass from dedicated perennial species to biofuels is developed, in the absence of government intervention, because of the potential efficiencies from coordinated harvest, storage, and delivery, market forces are likely to drive the system toward vertical integration. The structure of the industry is more likely to resemble global timber production, harvest, and delivery than atomistic grain production systems across the world.

Lastly, an innovative model was introduced that enabled determination of the tradeoff between the opportunity costs of closing the biorefinery as a result of insufficient feedstock and the quantity of contracted land considering both spatial and temporal bioenergy crop biomass yield variability. The model was used to determine the optimal quantity of land to lease as a function of the net cost of closing the biorefinery due to insufficient feedstock.

The value of considering spatial and temporal switchgrass yield variability when selecting land to lease will differ across regions. The models can be used to determine the value for any regions in the world.

References

Casler, M.D., Vogel, K.P., Taliaferro, C.M., Ehlke, N.J., Berdahl, J.D., Brummer, E.C., Kallenbach, R.L., West, C.P., Mitchell, R.B., 2007. Latitudinal and longitudinal adaptation of switchgrass populations. Crop Sci. 47, 2249–2260.

Kazi, F.K., Fortman, J.A., Anex, R.P., Hsu, D.D., Aden, A., Dutta, A., Kothandaraman, G., 2010. Techno-economic comparison of process technologies for biochemical ethanol production from corn stover. Fuel 89, S20–S28.

Lewis, S.M., Kelly, M., 2014. Mapping the potential for biofuel production on marginal lands: differences in definitions, data and models across scales. ISPRS Int. J. Geo-Informat. 3, 430–459.

Mondzozo, A.E., Swinton, S.M., Izaurralde, C.R., Monowitz, D.H., Zhang, X., 2011. Biomass supply from alternative cellulosic crops and crop residues: a spatially explicit bioeconomic modeling approach. Biomass Bioenergy 35, 4636–4647.

Richards, B.K., Stoof, C.R., Cary, I.J., Woodbury, P.B., 2014. Reporting on marginal lands for bioenergy feedstock production: a modest proposal. Bioenergy Res. 7, 1060–1062.

Sala, O.E., Parton, W.J., Joyce, L.A., Lauenroth, W.K., 1988. Primary production of the central grassland region of the United States. Ecology 69, 40–45.

Thomson, A.T., Izarrualde, R.C., West, T.O., Parrish, D.J., Tyler, D.D., Williams, J.R., 2009. Simulating Potential Switchgrass Production in the United States. Pacific Northwest National Laboratory. DE-AC05-76RL01830.

Williams, J.R., Jones, C.A., Dyke, P.T., 1984. A modeling approach to determining the relationship between erosion and land productivity. Trans. Am. Soc. Agric. Eng. 27, 129–144.

Wright, M.M., Daugaard, D.E., Satriob, J.A., Brown, R.C., 2010. Techno-economic comparison of biomass-to-biofuels pathways. Fuel 89-1, S2–S10.

CHAPTER 14

The role of biofuels in a 2 degrees scenario

Céline Giner*, Claire Palandri[†], Deepayan Debnath[‡]
[*]Trade and Agriculture Directorate, Organisation for Economic Co-operation and Development, Paris, France
[†]School of International and Public Affairs, Columbia University, New York, NY, United States
[‡]Food and Agricultural Policy Research Institute, University of Missouri, Columbia, MO, United States

Contents

Acronyms

2DS	Two degrees scenario
BAU	Business-as-usual
CO2e	Carbon dioxide-equivalent
EJ	Exajoule
ETP	Energy Technology Perspectives
FEW	Food, energy and water
GHG	Greenhouse gas
Gt	Gigatons
IEA	International Energy Agency
kt	Kiloton
LUC	Land use change
OECD	Organization for Economic Cooperation and Development
RFS	Renewable Fuel Standard
UNFCCC	United Nations Framework Convention on Climate Change
WTW	Well-to-wheels

Biofuels, Bioenergy and Food Security
https://doi.org/10.1016/B978-0-12-803954-0.00014-0

1. Introduction

The Paris Agreement, adopted in December 2015 at the COP21[1] and ratified by 180 Parties (as of August 2018), set at the forefront of the international mitigation agenda a common climate target. It set the objective of "holding the increase in the global average temperature to well below 2°C above pre-industrial levels and pursuing efforts to limit the temperature increase to 1.5°C above pre-industrial levels."

This 2 °C target, already alluded to by the scientific community since the 1980s, has now been well established as a critical limit for climate policy, above which disruptions to the climate system—and its impacts on other critical Earth-system processes—are likely to have dramatic and uncontrollable consequences.

How is this ambitious—yet necessary—target to be met? As the energy sector accounts for two-thirds of global greenhouse gas (GHG) emissions, it must play a major role in mitigation efforts. The Energy Technology Perspectives series published by the International Energy Agency (IEA) has focused on this target for several years, and developed a "2 degrees" scenario (2DS), which sets out a rapid decarbonization pathway in line with the international target.

The 2DS—furthered detailed in the next section—defines specific technology development pathways for the global energy system, and focuses on the transport, buildings, industry, and power sectors. In particular, it expects about 20% of the effort to be provided by the transport sector, with emissions peaking then declining sharply within the next decade. Bioenergy would provide 29% of the total transport final energy demand by 2060. Both the total volumes and the shares of specific categories of biofuels need to increase significantly.

In Chapter 8 we assessed the net historical contribution of biofuels to climate change mitigation up to 2017, then to 2027 in a business-as-usual (BAU) scenario, by using the partial equilibrium model Aglink-Cosimo. The model was adapted to estimate the impacts of biofuel use in terms of GHG emissions.

Building on this first assessment of the poor mitigation achieved until now and its concerning nonalignment with the goal adopted by the international community, this chapter goes one step further and tests the assumptions of biofuel requirements to meet the 2DS on the dynamics of agricultural markets. In particular, it answers the following critical question: Given the policies and intrinsic dynamics of the agricultural sector, is the development of biofuels as envisioned by the IEA in a 2DS indeed achievable with the incentives implemented (notably, increasing carbon tax levels on transportation fuels)? By explicitly considering the interconnection between agricultural and biofuel markets,

---

[1] 21st session of the Conference of the Parties to the United Nations Framework Convention on Climate Change.

the following sections explore the medium-term ability of agriculture to supply the amount of bioenergy deemed necessary to meet climate change targets in the transport sector.

This analysis also sets the stage for further prospective assessments of the contribution of biofuels up to the second part of the century, for which the 2DS trajectory requires an important transition from conventional to almost exclusively advanced biofuels.

2. The IEA's perspective on meeting the 2°C target and requirements for the biofuels sector

Which technology deployments and policy implementations in the energy sector are required to stay below a 2°C temperature increase by the end of the century? The IEA Energy Technology Perspectives report explores this question annually by laying out a development path for the energy sector in its central "2 degrees" mitigation scenario.

The 2DS corresponds to an energy system pathway and a CO_2 emissions trajectory "consistent with a chance to limit the mean global temperature increase below 2°C by 2100 of at least 50%." As such, it represents a highly ambitious transformation of the global energy sector that relies on a substantially strengthened response compared with today's efforts. It is built on strong assumptions of the types of measures and the deployment of energy technologies that would be required to attain this goal, and their implications for energy policy.

In the following sections we first describe the main assumptions and results of the 2DS and then focus on the specific role of biofuels.

2.1 Construction and main assumptions of the 2DS

The 2DS is rooted in the initial calculation of a "global carbon budget" for the given temperature target over a particular timeframe. This corresponds to the estimated amount of GHG emissions anthropogenic activities must not exceed to have a reasonable chance to remain under an average temperature increase of 2°C. After deducting the anticipated contribution from nonenergy sector GHG emissions, we get at the role of the energy sector in meeting the target. The cumulative energy sector CO_2 emission budget in the 2DS is thereby capped at 1170 gigatons of carbon dioxide ($GtCO_2$) over the period 2015–2100, reduced to 1000 $GtCO_2$ by the 2060 horizon.

This aggregate budget must also satisfy a specific pace of decarbonization over the period 2015–60 to be respected. The 2DS therefore requires emissions to peak before 2020 and to fall to around 25% of 2014 levels by 2060.

Delivering these cumulative emissions reductions requires a portfolio of technologies across sectors. The highest reduction levels are to be achieved through efficiency and reliance on renewable sources of energy, contributing a 40% and 35% share of the overall mitigation effort, respectively.

Importantly, the 2DS has the share of biomass and waste use in the energy mix double by 2060 to reach 22% in 2060. As the potential supply of biomass is constrained by the overall availability of sustainable feedstocks, the use of biomass across sectors is capped at 145 EJ. Bioenergy use is then directed primarily to sectors for which alternative decarbonization opportunities are limited; in particular, the transport sector.

2.2 Transport and biofuels in a 2 degrees world

The 2DS sets ambitious targets for the transport sector, which is to provide a fifth of the overall GHG abatement in the energy sector. Fifty-three percent of this effort would be achieved by higher energy efficiency, 41% through the use of renewables, and the remaining 6% by fuel switching.

This division of abatement effort is based on the adoption of an "Avoid, Shift, and Improve" strategy: first, a maximum of progress must be achieved by reducing energy consumption through decreasing activity, for example, through reducing demand for transport—both passenger and freight—or reducing trip length. Modal shift measures can further lower emissions by promoting transport modes that induce fewer emissions to achieve the same level of transport, and is thus the second-best abatement strategy. Finally, improvements on all modes of travel and technologies, from vehicle efficiency gains to higher substitution levels of fossil fuels by low-carbon-intensity biofuels, are needed to meet the 2DS objective.

Although a significant decrease in road energy consumption needs to occur to achieve the 2DS,[2] reductions in transport demand and energy use are not sufficient to stay within the 2°C carbon budget. A significant share of the effort needs therefore to come from biofuels, especially for the long-haul modes of travel and freight that have few alternatives to fuel, such as aviation and maritime transport. The 2DS therefore sees a major expansion of the role of bioenergy in the transport sector, reaching nearly 30 EJ in 2060 (2.5 times that expected in a BAU scenario, and nearly 10 times the level in 2016).[3]

Fig. 14.1 presents the relative contribution of the different types of biofuels (liquid and gaseous) by 2060. Two main features stand out. First, meeting the 2DS target requires a huge rise in biofuel production. This may only be possible with a supporting economic environment. The 2DS therefore relies on an ambitious carbon pricing mechanism with

---

[2] "Avoid" measures for passenger transport result in a 25%–27% reduction in passenger activity (measured in passenger kilometers) for cars by 2060 relative to the baseline. Vehicle kilometers traveled by road freight are also reduced by 16%–26%.

[3] These optimized effort levels result from simulations undertaken with the IEA's Mobility Model, which enables detailed projections of transport activity, vehicle activity, energy demand, and well-to-wheel (WTW) GHG emissions according to policy scenarios up to 2060.

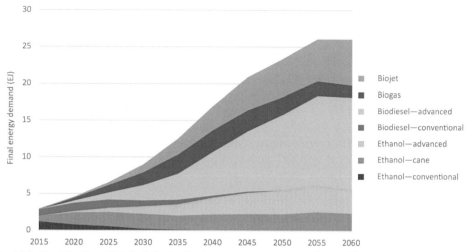

Fig. 14.1 Contribution of biofuels in the final transportation energy demand by 2060 in the 2° scenario. *(From OECD/IEA Energy Technology Perspectives, 2017. Catalysing Energy Technology Transformations. IEA, Paris. https://doi.org/10.1787/energy_tech-2017-en, p. 325).*

carbon taxes on fossil fuels that reflect lifecycle GHG emissions intensities continually increasing over the entire period.

Second, the nature of the biofuels pool changes considerably over the 45-year period, with notably the gradual disappearance of conventional biodiesel and concomitant surge of advanced biodiesel. The levels of advanced ethanol also increase markedly and exceed that of conventional ethanol by the middle of the period. Supply of conventional ethanol is stable; however, its composition changes significantly as sugarcane-based ethanol gradually replaces other feedstocks and constitutes virtually the entire supply volume by 2035. Indeed, on average, ethanol produced from sugarcane has a stronger GHG reduction profile than when produced from other agricultural crops[4] (see Figure 1 of Chapter 8). It thus offers the highest GHG emission savings of the conventional pool and is ensured a continuing role in abatement efforts thanks to its relatively low production costs.

The distinction between "conventional" and "advanced" categories and what these comprehend is largely discussed in Chapter 2. In the context of the 2DS, advanced biofuels include fuels produced from nonfood crop feedstocks and which have high GHG emission savings potentials (e.g., agricultural and forestry residues and waste oils and fats).

---

[4] It is important to note that this lower carbon intensity of sugarcane ethanol is mostly due to the use of sugarcane bagasse in place of the usual coal during the conversion process. While this is the case in Brazil—which represented 23% of global ethanol production in 2017—the transformation process in other countries is still based on the use of coal, which leads to a carbon intensity of the final biofuel of the same order of magnitude of other sugar crop-based types of ethanol.

The results of this new transport fuels mix are a decrease in WTW emissions to 4.4 Gt in 2060.

The 2DS therefore sees an important role for bioenergy in decarbonizing the energy sector to meet the global climate change mitigation goal, and in particular for biofuels in the transport sector. This raises the key question of the extent to which these desired levels and trajectory are reconcilable with the dynamics of agricultural markets.

3. Are the 2DS biofuel volumes compatible with the dynamics of the agricultural sector and global agricultural markets?

To answer this question, we bring together the results of the IEA's energetic model and a global model of main agricultural markets, to allow for an improved consideration of biofuels, as they lie at the intersection of the two sectors. We will focus in particular on the developments needed in the medium term. As seen in Fig. 14.1, by 2030 the global supply of biofuel will still largely rely on conventional feedstocks, as more advanced technologies are slowly scaling up. As such, much of the effort to be supplied during that period is inherently interconnected with the forces at play on agricultural markets.

3.1 Aglink-Cosimo: A partial equilibrium model of world agricultural markets

Aglink-Cosimo is a partial equilibrium model of world agricultural markets, with a detailed production, use, and trade modeling framework for most categories of biofuels currently available on the market and directly connected to agricultural markets. Markets outside of the agricultural scope are not explicitly represented and major macroeconomic variables such as exchange rates are entered as exogenous inputs. Importantly, in addition to projecting a baseline of market conditions in a 10-year horizon, this model provides the structure to run scenarios.

Aglink-Cosimo is a recursive model, i.e., values of key economic variables from each year are dependent on values from past years. For each commodity considered, a world price clears the world market, ensuring that global net trade is null.

Aglink-Cosimo is built on insights into policies implemented in each country represented. This means that although a common framework is used between countries—at least among countries in a same module—equations are adapted to account for specific policies in these countries. Aglink-Cosimo distinguishes two commodities: biodiesel and ethanol.

To have a general understanding of how key variables related to biofuels react to market changes, the general structure for the representation of biofuels markets in the model is detailed in Box 14.1 for the example of ethanol. For each country considered, this general structure can be adjusted to account for specific agricultural policies. The biofuel component of the Aglink–Cosimo model is described in detail in OECD (2018), and more details on the overall structure of the model can be found in OECD (2015).

BOX 14.1 Representation of biofuels markets in Aglink-Cosimo; example of ethanol

- A market clearing price closes the balance at the global level: WLD_ET_XP
- A second domestic market clearing price ET_PP closes the balance at the national level:

$$0 = ET\_QP_i + ET\_IM_i - ET\_QC_i - ET\_EX_i$$

- On the supply side: the total production of ethanol is the sum of the production from each possible feedstock. The production function from each feedstock has a price reaction component (i.e., QP reacts on the price difference between ethanol and the feedstock) and a proxy for production capacity (i.e., QP reacts on the development of the crop). The following example corresponds to the production of ethanol from sugarcane:

$$ET\_QP = \sum_i ET\_QP_{feedstock-i}$$

$$\log(ET\_QP..SCA) = \alpha + \beta_1 \times \log\left(\frac{ET\_PP}{SCA\_PP}\right) + \beta_2 \times \log\left(ET\_QP..SCA_{(-1)}\right) + \log(R)$$

- On the demand side: the fuel use of ethanol is determined by the product of the consumption of gasoline and the share of ethanol (divided by the energy content ratio between ethanol and gasoline: about 0.67). This "effective" share is itself defined as the maximum value between the theoretical share (market driven) and the blending obligation (mandate driven). Finally, this market-driven share reacts on the price difference between ethanol and gasoline. When the relative consumer price of gasoline compared to that of ethanol increases, i.e., when ethanol becomes more competitive, the market-driven share increases.

$$ET_{QC} = ET_{FL} + ET_{OU}$$

with :

$$ET_{FL} = GAS_{QC} \times ET_{QCS}..EFF \div 0.67$$

$$\log(ET_{OU}) = \alpha + \beta_1 \times \log\left(\frac{ET_{PP}}{CPI}\right) + \beta_2 \times \log(GDP) + \log(R)$$

$$ET_{QCS}..EFF = \max(ET\_QCS, ET\_QCS..OBL)$$

$$\log(ET_{QCS}) = \alpha + \beta \times \log\left(\frac{ET_{CP}}{GAS_{CP}}\right) + \log(R)$$

Continued

BOX 14.1 Representation of biofuels markets in Aglink-Cosimo; example of ethanol—cont'd

- Trade:

$$\log(ET\_IM) = \alpha + \beta \times \log\left(\frac{ET\_PP}{ET\_IMP \times (1 + ET\_TAVI)}\right) + \log(R)$$

$$\log(ET\_EX) = \alpha + \beta \times \log\left(\frac{ET\_PP}{ET\_EXP \times (1 + ET\_TAVE)}\right) + \log(R)$$

where the variables' names are denoted using the following structure: Commodity_Item, with eventually an additional dimension for details on the item: Commodity_Item…Detail. In the equations above, we used the following names:

- Commodities: ET = ethanol; GAS = gasoline; SCA = sugarcane
- Items: EX = exports; EXP = export price; FL = fuel use; IM = imports; IMP = import price; OU = other uses; PP = producer price; QC = consumption; QCS = consumption share; QCS..EFF = effective consumption share; QCS..OBL = blending obligation share; QP = production; QP..SCA = production from sugarcane; TAVE = export tax (in %); TAVI = import tariff (in %); XP = world reference price (US dollars/unit of measure)
- Others: CPI = consumer price index; GDP = growth domestic product; R = residual.

For example, ET_QCS..OBL represents the blending obligation share of ethanol in the gasoline pool, expressed in %.

All prices are expressed in domestic currency.

Two noteworthy specificities of the model are that the different feedstocks of origin are represented only on the supply side; on the demand side, only the two "ethanol" and "biodiesel" aggregates remain. In addition, the model does not include a spatial representation of trade but "assumes homogeneity on the world market for all commodities," such that, for example, imports of a commodity are independent of its country of origin (OECD, 2015). However, Chapter 8 showed how the carbon intensity can greatly vary across feedstocks as well as conversion technologies, which can be proxied by the producer country. Estimations of the emissions associated with the consumption of biofuels must therefore consider the original feedstocks from which they were produced. To account for this, one can conduct a short sequence of steps to "match" volumes of biofuel consumption to the raw material by estimating trade fluxes and reallocating biofuels exchanges by country—and thereby feedstocks—of origin. These are detailed in Box 14.2 for information purposes but are not the main focus of this chapter.

BOX 14.2 Module to allocate biofuel GHG emissions to consumer countries; example of ethanol

The mnemonics used in the following equations are identical to those of Box 14.1. $f_i - Et$ refers to category i of feedstocks used for ethanol production.

1. On the supply side, the production of each feedstock category identified is represented for each producer country.

2. Exports: We make the simplifying assumption that the split of a country's exports is identical to the split of its production. For each country, we estimate the volume exported for each of the I feedstocks considered:

$$EX_{f_i - Et} = \frac{QP_{f_i - Et}}{\sum_{i=1}^{I} QP_{f_i - Et}} \times EX_{Et}$$

3. We aggregate the volumes exported to obtain the global volume exported of each feedstock.

4. Imports: The calculation steps follow as if: first, all N countries proceed to their respective exports of ethanol, together adding to a global volume available on the market; then importing countries proceed to their respective imports. To deal with missing information on the type of ethanol imported by some countries, we proceed as follows: for the major importing countries, we determine import shares from internal data and feed them exogenously to the module. Their imports of ethanol from each feedstock are subtracted from the global volumes available on the market. The remaining k countries then import their respective volumes of ethanol from each feedstock according to the split of the remaining volume on the market:

$$IM_{f_i - Et} = \frac{Global^k IM_{f_i - Et}}{Global^k IM_{Et}} \times IM_{Et}$$

5. The consumption of ethanol from each feedstock is then calculated for each country through the following formula:

$$QC_{f_i - Et} = QP_{f_i - Et} + IM_{f_i - Et} - EX_{f_i - Et}$$

Note: There is a specificity regarding the ethanol module, because ethanol is not used solely as a transportation fuel. Other end uses include industrial chemicals and alcoholic beverages. Total consumption and fuel use are thus distinguished on the demand side. For the purpose of our analysis, the share of ethanol used for transport purposes is therefore first derived from total ethanol use.

6. Finally, GHG emissions in $ktCO_2e$ are estimated for each country by matching the consumption of biodiesel from each feedstock with its emission factor:

$$GHG\,emissions_{Et} = \sum_{i=1}^{I} QC_{f_i - Et} \times emission\ factor_{f_i - Et}$$

3.2 Definition of an Aglink-Cosimo scenario

The following analysis leverages the capacity of the modeling framework to ask the following question: Given the policies and intrinsic dynamics of the agricultural sector, is the medium-term evolution of biofuels as envisioned by the IEA in a 2DS indeed achievable through the incentives implemented (notably, increasing carbon tax levels on transportation fuels)?

After calibration of the model—necessary to maintain consistency with the assumptions of the 2DS regarding the future evolution of crude oil prices, key macroeconomic variables, taxes applied to fuels according to their GHG profiles, and the growth rate of demand for gasoline and diesel—the 2DS can be "tested" on the actual dynamics of agricultural markets. Table 14.1 summarizes the set of assumptions involved, and compares them with the conditions in the baseline (i.e., what is expected on agricultural markets in a BAU scenario where current policy settings remained unchanged).

Table 14.1 Main assumptions in the calibration of Aglink-Cosimo to the IEA's 2DS

		Adapted 2DS scenario			Baseline	Differences w.r.t. baseline
		2020	2025	2030	2030	2030
Crude oil prices	USD/barrel	87.6	111.9	140.1	79.7	76%
Additional taxes applied to fuels						
Expressed in terms of carbon equivalent						
Gasoline-type fuels	USD/tC02e	24.4	54.6	91.7	..	..
Diesel-type fuels	USD/tC02e	18.5	41.5	69.8	..	..
Expressed as a WTW emission based tax						
Gasoline	USD/hL	6.5	14.6	24.5	..	..
Diesel	USD/hL	6.6	14.8	24.9	..	..
Sugarcane based ethanol	USD/hL	1.3	2.9	4.9	..	..
Maize based ethanol	USD/hL	3.3	7.5	12.5	..	..
Ag. Residues based ethanol	USD/hL	0.7	1.5	2.5	..	..
Palm oil based biodiesel	USD/hL	4.0	8.9	14.9	..	..

Table 14.1 Main assumptions in the calibration of Aglink-Cosimo to the IEA's 2DS—cont'd

		Adapted 2DS scenario			Baseline	Differences w.r.t. baseline
		2020	2025	2030	2030	2030
Crude oil prices	USD/barrel	87.6	111.9	140.1	79.7	76%
Soybean oil based biodiesel	USD/hL	3.8	8.5	14.2	..	..
Rapeseed oil based biodiesel	USD/hL	3.4	7.5	12.7	..	..
Waste oil based biodiesel	USD/hL	1.2	2.6	4.3	..	..
Demand In key countries						
Gasoline-type fuels						
World	bln L	1268	1140	998	1318	−24%
USA	bln L	542	455	367	454	−19%
European Union	bln L	117	94	76	103	−26%
Brazil	bln L	47	47	45	51	−11%
China	bln L	173	176	164	229	−28%
India	bln L	37	42	48	87	−45%
Diesel-type fuels						
World	bln L	996	1021	1024	1047	−2%
USA	bln L	217	204	186	199	−7%
European Union	bln L	229	209	186	204	−9%
Brazil	bln L	52	53	54	56	−3%
China	bln L	125	132	134	126	6%
Indonesia	bln L	38	46	52	52	0%
Argentina	bln L	13	14	14	16	−13%

Note: An additional assumption is made to increase the ethanol blend wall to 15% to allow for more ethanol to be blended with gasoline, similarly to the assumptions of the IEA 2DS. *USD/hL*, US dollars/hectoliters; *WTW*, well to wheels.

In the footsteps of the original 2DS, carbon taxes are affected to very conservative emission factors for biofuels, in that they only consider WTW emissions and exclude those stemming from land-use changes (LUCs). The composition of emissions factors of biofuels and the debate around the inclusion of LUC-related emissions is discussed in detail in Chapter 8.

3.3 Results: Implications for biofuels and agricultural markets up to 2030

When confronted with the response dynamics of agricultural markets, the development of biofuel blending in transportation fuels is less pronounced than in the IEA 2DS. The 2DS expects a strong development of sugarcane-based ethanol production in the period leading to 2025 (+130% at the expense of maize-based ethanol production) followed by an uptake of advanced ethanol production. For biodiesel, the IEA 2DS sees a substantial increase in vegetable oil-based biodiesel up to 2025 (+39%) and then the take-off of waste oil-based biodiesel. The results of the simulation of the agricultural model show a much lower biofuel use growth over the period leading to 2030, suggesting that the economic environment and policies implemented are not favorable to such a rise in biofuel consumption.

Indeed, in Aglink-Cosimo, biofuel use is promoted by the taxes applied to fuels according to their GHG emission profiles. However, the production of conventional biofuels is constrained by the availability of agricultural feedstock that can be supplied by agricultural markets in addition to food and feed demand. A doubling of the consumption of ethanol produced from sugarcane would imply a very strong shock on sugar markets. On biodiesel markets, however, waste oil-based biodiesel, however, can increase substantially because it is not competing with food demand, and production reaches levels 27% higher than in the baseline (BAU scenario) at the expense of vegetable oil-based biodiesel.

In all countries, ethanol and biodiesel blending shares in respectively gasoline- and diesel- type fuels are stronger than in the baseline because the taxes applied to the different fuels according to their WTW GHG profiles decrease the price ratio between biofuels and conventional fuels, thus encouraging the market-driven use of biofuels behind mandates (Fig. 14.2). This market-driven effect is particularly strong in China where the ethanol share in gasoline-type fuels doubles to reach 4% by 2030. In Brazil, it is the use of hydrous ethanol (pure ethanol that can be used by the fleet of flex-fuel cars) that strongly increases as a response to tax stimulus.

Most strikingly, although the overall volumes traded remain relatively stable with respect to the baseline, the share of biofuels with lower GHG emission profiles in total biofuel trade increases strongly (+25% for sugarcane-based ethanol and +90% for waste oil-based biodiesel).

The mitigation efforts achieved in the 2DS are dependent on a set of assumptions, not the least of which are the global crude oil prices (Table 14.1). By 2030, their levels are assumed to be almost twice as high as in the baseline. However, higher oil prices increase agricultural production costs through higher prices for fuel and fertilizer—as well as through general cost increases induced by higher inflation—and therefore can highly affect the demand for agricultural commodities. These indirect effects are picked up by the Aglink-Cosimo model, and induce, for example, a stronger consumer food price index by about 1.4% when compared to

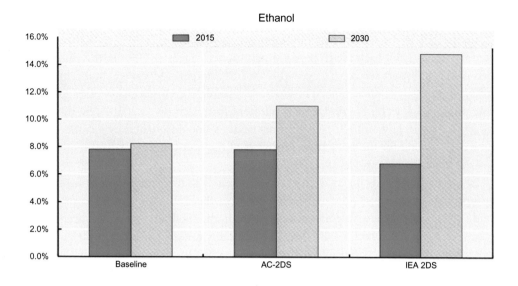

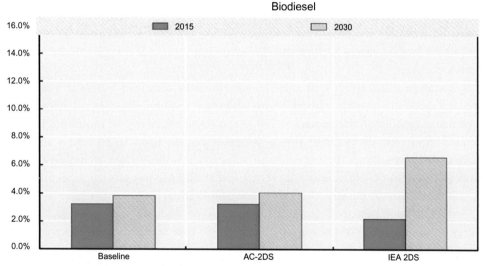

Fig. 14.2 Comparison of biofuel blending shares in volume in 2015 and 2030. *Note*: 2015 data used in the International Energy Agency scenario are estimates. *AC-2DS*, Aglink-Cosimo-2DS; *IEA 2DS*, International Energy Agency 2° scenario. *(Source: OECD Secretariat based on Aglink-Cosimo simulations and OECD/IEA (2017)).*

the baseline. Aside from these, overall effects of the 2DS on agricultural markets are relatively moderate. Indeed, because of the lower demand for gasoline and diesel in the medium term, the increase in the share of biofuel in total fuel use does not lead to a strong increase in demand for agricultural feedstock for biofuel production.

3.4 Complying with the 2DS trajectory to 2060—The transition to advanced biofuels from agricultural residues

Let us now look at the second part of the transition period: from 2030 to 2060. As displayed in Fig 14.1, biofuels are to play a much higher role after 2030 in the mitigation efforts of the transport sector to stay within the 2°C target. However, these important increases stem solely from advanced biofuels. As a result, the composition of the biofuels supply pool in this second period is very different from that up to 2030. The transition period from a virtually entirely conventional pool to one dominated by advanced fuels will be determinant for the mitigation prospects of the transport sector, and requires the following issues to be explored thoroughly.

The production of advanced biofuels remains largely in the experimental phase. If the scale-up of the industry is slower than expected in the 2DS, and the transport sector would thereby rely mostly on conventional crops in the medium term, then accounting for the—direct and indirect—LUC impacts and their related emissions is all the more important in the context of ambitious mitigation goals. In addition, if biofuel markets do not see a continuation of the production levels of the conventional crop-based feedstocks, but actually an increase in their supply (for example, due to high blending mandates not conditioned on biofuels not competing with agricultural crops, nor accompanied by sufficient restrictions on their conversion process or carbon intensity), what this implies both in terms of net emissions savings and impacts on agricultural markets needs to be assessed.

To analyze the feasibility of the "conventional-to-advanced biofuels" transition, three key questions must then be addressed, and more research is needed on these topics: How many agricultural residues are available and could effectively be mobilized to meet the 2DS required levels by 2060? Which policies—notably financial incentives and government support—would be necessary to achieve the scale-up of the industry in the required timeframe? Which environmental and economic impacts might result from the transformation of the structure of the biofuels industry?

The production cost of ethanol from agricultural residues, forest residues, and specific energy crops will be a determining factor of both the pace and the scale of the transition. However, as current production technology is not yet available on a large scale (OECD/IEA, 2017), information on the production costs of technologies based on cellulosic conversion are not widely available. For new policies and regulations to be most conducive to the scale-up of the industry, these must be informed by the shape of the cost function. The extent to which the usual learning curve[5] assumption is relevant in this context needs

---

[5] Production costs and learning curves: To represent the evolution of technology costs, models often rely on a "learning curve," which reflects the assumption that the efficiency of production (and therefore the reduction of cost) increases with experience. More specifically, production cost can be assumed to be brought down by a specific ratio when the number of units produced (in our case the volume of biofuel from agricultural residues) is doubled. While the slope of the curve is very steep with the first units produced, it gradually becomes flatter and is finally asymptotic.

BOX 14.3 Current investments and policies on advanced biofuels from lignocellulosic material

Many governments have launched and supported research and development programs since the 2000s on various categories of advanced biofuels. Several projects have developed past the experimental stage to reach commercial scale. However, they still represent only a very small percentage of global biofuel production. Moreover, the operating rates in many current production plants have proved to be considerably below capacity. Multiple plants also abandoned production after a few years due to distribution problems, lack of infrastructure, or because they weren't competitive enough, and several other projects were canceled.

Targets set by several governments are still insufficient to boost the development of the advanced biofuels sector. The European Union had set a low indicative target of 0.5% share of energy content to be transposed by member states by 2017. In Finland, for instance, the National Climate and Energy Strategy adopted in 2013 aims at increasing the production of biofuels and emphasizes on advanced biofuels and the use of waste and by-products. However, it does not provide any quantitative or precise target, and reflects the current concerns and "uncertainties" created by the European Union's decision to limit the use of food-based first-generation biofuels. A few more rigorous policies have started to develop: in 2015, Italy became the first country to impose blending requirements for advanced biofuels at 1.2% by 2018. The United States has also set volume requirements for "advanced biofuels"; however, the RFS definition includes a vast range of categories of biofuels, notably Brazilian sugarcane-based ethanol. The lack of a globally recognized definition of advanced biofuels is also detrimental to the scaling-up of the industry.

---

to be confronted with insights from the industry on technology development and production capacities (Box 14.3).

Furthermore, the full production cost of ethanol or renewable diesel from agricultural residues comprises not only conversion, transport, and collecting costs, but also an "opportunity cost," i.e., the fertilizer value and soil quality value of biofuels. This opportunity cost may be substantial. In addition, leveraging the potential of agricultural residues for biofuel production implies shifting these resources away from their current uses: lignocellulosic residues may be burnt to generate heat or electricity for buildings and industry. Given the potential limited availability of this biomass, the IEA underlines that bioenergy use "must be concentrated in sectors for which alternative decarbonization opportunities are limited," such as various modes of transportation. To assess both the feasibility and the expected implications of this transition to advanced biofuels from lignocellulosic material, more research on their opportunity costs and potential adverse effects on other energy sectors is needed.

4. Conclusion

Confronting the biofuel volumes requirements for abatement in the transport sector with the dynamics of agricultural markets revealed factors that may impede the required developments on biofuels markets. The results showed that under the policies and instruments contemplated, the evolution of agricultural production costs combined with the constrained availability of the feedstock are not able to induce the biofuel blending shares projected in the IEA 2DS. By 2030, the supply of biofuels, even with conservative emission factors, is not in alignment with the objectives of the Paris Agreement as modeled through a 2DS.

In sectors for which limited low-carbon alternatives exist, bioenergy needs, however, to play an important role in the portfolio of measures and technologies required for the decarbonization of energy sectors. The pace of the phaseout of conventional feedstocks to advanced biofuels is one of the key challenges for the transport sector in the decades ahead.

This analysis also showed that the development of the biofuel industry is embedded in important interdependencies of the food and energy sectors. Another key component of this nexus, which should not be underestimated, is water. The concept of integrated Food, Energy and Water systems has gained prominence in recent years because the interdependencies among and tradeoffs between these systems are becoming both increasingly apparent and tight.

These interdependencies on multiple levels make examining food, energy, and water systems jointly highly informative, and reveal substantial efficiencies that can be missed when systems are optimized individually (Lant et al., 2018). We saw here the importance of taking a multisectoral approach to study the supply and demand of goods positioned at the intersection of these sectors. Enlarging this approach to include interactions with the water sector is an important area for future research.

Finally, taking fully into account land-use impacts of not only conventional but also advanced types of biofuels should not be neglected even as the industry changes. Land use transforms not only the balance of GHG emissions but also Earth's terrestrial surface, resulting in changes in key biogeochemical cycles and the ability of ecosystems to deliver critical services (Haberl et al., 2007). The potential adverse effects of biomass harvest on ecosystems needs to be analyzed more deeply, especially in the context of a scheme to substitute biomass for fossil fuels on such a massive and global scale.

Disclaimer

The views expressed are those of the author and do not necessarily represent the official views of the OECD or of the governments of its member countries.

References

Haberl, H., Erb, K.H., Krausmann, F., Gaube, V., Bondeau, A., Plutzar, C., Gingrich, S., Lucht, W., Fischer-Kowalski, M., 2007. Quantifying and mapping the human appropriation of net primary production in earth's terrestrial ecosystems. Proc. Natl. Acad. Sci. U. S. A. 104 (31), 12942–12947.

Lant, C., Baggio, J., Konar, M., Mejia, A., Ruddell, B., Rushforth, R., Sabo, J.L., Troy, T.J., July 2018. The U.S. food-energy-water system: a blueprint to fill the mesoscale gap for science and decision-making. Ambio.

OECD, 2015. Aglink-Cosimo Model Documentation—A Partial Equilibrium Model of World Agricultural Markets. Technical report, OECD.

OECD, April 2018. Biofuel Module Documentation. Accessed online, http://www.agri-outlook.org/about/Aglink-Cosimo-Biofuel-Documentation.pdf.

OECD/IEA, 2017. Energy Technology Perspectives 2017: Catalysing Energy Technology Transformations. IEA, Paris. https://doi.org/10.1787/energy_tech-2017-en.

Further reading

BioDieNet, 2009. El LIBRO—The Handbook for Local Initiatives for Biodiesel From Recycled Oil. Technical report, IEE Programme (Intelligent Energy Europe), EIE/06/090.

Hillairet, F., Allemandou, V., Golab, K., 2016. Analysis of the Current Development of Household UCO Collection Systems in the EU. Technical report, GREENEA.

OECD, August 2008. Biofuel Support Policies: An Economic Assessment. OECD Publishing.

Spöttle, M., Alberici, S., Toop, G., Peters, D., Gamba, L., Ping, S., Steen, H., Belle-fleur, D., 2013. Low ILUC potential of wastes and residues for biofuels: straw, forestry residues, UCO, corn cobs. Ecofys, Utrecht Google Scholar.

Toop, G., Alberici, S., Spöttle, M., Van Steen, H., 2013. Trends in the UCO Market. Ecofys, London.

U.S. Department of Agriculture, Foreign Agricultural Service (FAS), August 2015. Canada Biofuels Annual, GAIN Report Number: CA15076. Technical Report.

Index

Note: Page numbers followed by *f* indicate figures, *t* indicate tables, *b* indicate boxes, and *np* indicate footnotes.

Printed in the United States
By Bookmasters